THE NEW GRAVITY

IS IT A QUIRK OF NATURE TO CONFUSE MAN
OR DOES MAN UNWITTINGLY CONFUSE NATURE

THE NEW GRAVITY

A New Force · A New Mass · A New Acceleration

Unifying Gravity with Light

Kenneth G. Salem

Salem Books

Johnstown, PA

THE NEW GRAVITY, First Edition

Salem, Kenneth G.
The New Gravity

ISBN 0-9625398-1-3

This book I dedicate to my dear wife
Jean
whom without her undying support
I could not have done these works

Contents

Preface

This book, THE NEW GRAVITY, pertains not only to gravity per se; it pertains also to the GUTs and TOEs of physics—GUTs for Grand Unified Theories and TOEs for Theory of Everything—the ultimate goal of physicists around the world.

What physicists are pursuing above all is the connection that would unify the Gravitational interaction with the Electromagnetic interaction. That answer alone, it is believed, would be the beginning of the unraveling of many unexplained phenomena of nature.

It is the goal of The New Gravity to identify with as much detail as possible the SINGLE FORCE which is believed to be the responsible agent for creating both the Gravitational interaction and the Electromagnetic interaction.

In the process, The New Gravity will introduce a new perspective on the fundamentals of gravity and gravitation and will present a more direct and less complex method of comprehending gravity and how the force creating it relates, in particular, to the propagation and precise velocity of light.

The New Gravity will disclose several new principles which have naturally arisen as a result of having *redefined* a body's *Gravitational Mass* to be substantially different from what it has classically been known to be.

It has been believed since Newton's time that both Gravitational Mass and Inertial Mass are equivalent to one another. This does not, however, actually appear to be the case.

The New Gravity will instead show that, wherein *inertial* mass reflects a body's *total* mass, a body's *gravitational* mass will herein be redefined to reflect—*not* its total mass as has generally been believed—but *only* that product (solely) of its mean *density* and mean *radius* times *one-square meter*.

That specific formulation theoretically translates into a one-square meter column of gravitational mass which, for *each* kilogram of mass calculable from such column, there will be an acceleration near a spherical body's surface of 2.8 Angstroms (2.8×10^{-10} meter) per second per second for any freely falling body. In other words, it will be shown that the acceleration of *any* freely falling body due to gravity near a body's surface is directly proportional to the theoretical mean mass of a spherical body's Square-Meter Column, i.e., the distance from the attracting body's center of mass to any point of attraction whose length is always that of its radius.

In *essence* then, gravitational field strength g is directly proportional *"solely"* to the product of a body's mean radius and mean density; and the gravitational force of attraction F between any two bodies is simply the *product* of one body's gravitational *field strength* out from its center of mass to the attracted body's center of mass, and the attracted body's *total mass*.

It is, in fact, the constant of proportionality between gravitational field strength in the unit of Newton only, and column-mass in the unit of kilogram, which translates into an *acceleration* in strict accordance with Newton's Second Law of Motion, a = F/m. This in turn shows that since there exists a constant ratio of 2.8×10^{-10}ms^{-2} between *any* body's gravitational field strength and its column-mass, that all bodies must themselves be *eternally accelerating* thru space at a constant rate of 2.8 Angstroms per second per second (Note: an Angstrom unit is 10^{-10} meter).

It is that specific acceleration thru space which would appear to be what is actually creating the force of gravity one experiences at the surface of the Earth, or for that matter, at the surface of any solar system body.

And so it is, after all is said and done, no coincidence that the effect one feels from the force of gravity holding them down to the Earth's surface like a magnet, and the effect one would hypothetically feel inside of a rocket *accelerating* upwards in a gravity free environment at the Earth's gravity rate of 9.8 meters per second per second, that these two phenomena are indeed one and the *same* effect impossible to distinguish one from the other.

This effect—the gravity we experience at the Earth's surface—will be shown to be due to some as yet unidentified, but equivalent, Outside Force of 2.8×10^{-10} Newton acting on each and every body in the universe in strict proportion to each body's *column-mass*. So that once again, if we are using the Earth as an example, its gravitational mass, that is, its 35-billion kilograms calculable from its theoretical square-meter column, will cause it to experience a total Outside Force of 9.8 Newtons acting upon it, or otherwise pushing it thru space. The Moon's 5.8-billion kilograms of gravitational mass calculable from *its* column, will likewise cause *it* to experience a total Outside Force of 1.6 Newtons.

It will be seen then that Newton's Third Law of "equal but opposite action and reaction" is applicable in the two cases just described. For example, the 9.8 Newton Outside Force acting upon or pushing the Earth thru space—causing it to accelerate 2.8 Angstroms per second per second—is precisely what is causing the equal but opposite force of 9.8 Newtons (per unit mass) which we experience at its surface.

The Moon likewise, which has a total Outside Force of 1.6 Newtons acting upon or pushing it thru space with a constant acceleration of 2.8 Angstroms per second per second, also has that same equivalent equal but opposite force of 1.6 Newtons (per unit mass) pulling downward at any object resting upon *its* surface.

This then is what this book, THE NEW GRAVITY is all about. It will describe in detail those main points already brought out plus several other aspects which vary in one way or another from Newton's classical law of gravitation.

The most significant revelation about Newton's Universal Law of Gravitation is what is referred to as the $1/r^2$ phenomenon. This in effect, means for example, that if the centers of mass of two 1-kilogram spherical bodies were separated by one-meter, each body would impose a gravitational force upon the other of 6.672×10^{-11} Newton—exactly one unit of Newton's Universal gravitational constant, G. However, if those same two one-kilogram masses were separated by twice their original distance to two meters, then the Force between them would only be one-fourth ($6.672 \times 10^{-11} \div 2^2$) of the original force, otherwise, 1.668×10^{-11} Newton; and if separated by three times their original distance, the force of attraction would diminish to one-ninth ($6.672 \times 10^{-11} \div 3^2$) their original attraction, otherwise, 7.413×10^{-12} Newton.

This important law of $1/r^2$ is *not* lost in any way with the New Gravity. It remains intact just as Newton had originally described it to be; i.e., the force drops off as the inverse square of the distance. In fact, the New Gravity co-operates very well with Newton's classical gravity. However, there are several new principles and differences to be presented which make gravity and its *concept* much more comprehensible.

What is even more significant than the improvements the New Gravity makes to Newton's classical law of gravitation is the fact that the New Gravity's Universal Constant of Acceleration, $2.8 \times 10^{-10} \text{ms}^{-2}$, co-operates extremely well with other areas of physics where Newton's Universal Constant of Gravitation, 6.672×10^{-11} $\text{m}^3\text{kg}^{-1}\text{s}^{-2}$, does not.

Those areas spoken of include—but are not limited to—the speed of light; the age of the Universe; Hubble's constant in which it will be shown that this is, in essence, the same value as the *ratio* between a body's gravitational field strength and its square-meter column-mass, which otherwise is the New Gravity's Universal Constant of Acceleration, $2.8 \times 10^{-10} \text{ms}^{-2}$.

Although The first part of this book will deal directly with the New Gravity and its advantages over classical gravitation, the second part will deal with Unification matters which stem from the New Gravity, especially on the Unification of the Gravitational interaction and the Electromagnetic interaction, revealing especially, and with much detail, the manner in which light propagates and why it has the precise velocity it presently has.

Introduction

The premise of this book is conceived in a progression of thought over a period of forty-seven years. It concerns particular questions about the nature and speed of light, and leads to a systematic explanation of why the universe works the way it does. In this book, THE NEW GRAVITY, I will put forward evidence that some Outside Force is *eternally* driving each and every particle in the universe, thereby causing them to constantly accelerate at a rate of 2.8 Angstroms per second per second, creating a *counter* force we know as *gravity*.

While modern philosophers and scientists seldom do work in each others disciplines, I found no need to make a sharp distinction between the two. The questions I would ask could concern epistemology or cosmology. The main question, however, was always this: "Why does light travel at the speed that it does and why is that speed always constant?" The mystery of light's constant velocity

1

was not in that it, as Einstein said, was constant at any time and any place throughout all time and space, but that it never had to reach that velocity by starting first from an initial velocity of zero. The question, stated more succinctly, is: "Why does a light beam, say from a flashlight, start from the flashlight at 186,000 amps and not start from zero velocity *accelerating* up to 186,000?" This and other questions about light was a fascinating subject which I was always engrossed in. It was simply amazing to me that light or *anything* could, in fact, travel that fast, yet *start* at a real velocity and *not* from rest. It's like asking: "How can anything, or anyone, go on a journey without first having to start out?

I've always felt that certain key physical quantities of our world should not be taken for granted, and that it was important to have an intuitive grasp of what a physical quantity was indicating. Henceforth, I was always strict about using real numbers when tackling new equations because I wanted to know the quantities of physical constants and parameters. I believe this method builds a better foundation and understanding of the equations and their physical concepts. High school and college physics texts tend not to do enough of this, making it more difficult for students to grasp necessary concepts. This method led me to question things I may never have questioned in various areas of physics. The most compelling questions, though, concerned the initial and final velocity of light; and what relation light might have to gravity.

While I had questions about the kinetics of light, I was not disputing that light traveled at the universally accepted velocity of 186,000 miles per second. As James Clerk Maxwell said in his *The Nature of Light:* "that light always propagated with *one* specific velocity of 186,000 miles per second," Einstein agreed and extended this by saying light's velocity was not only constant, but was an *absolute and unchanging constant of nature throughout all space and time.* Before Einstein had made that statement, Michelson and Morley's experiments of the 1880s sought to detect a variation in the velocity of light due to the Earth's 18 mile per second orbit around the Sun. However, their experiments failed to confirm that light behaved according to the principle of the addition of velocities as does sound. It was after that (several years later) in which Einstein declared his views on the subject in which he also dispelled the suggestions that an *ether* existed in order to carry the light waves.

Michelson and Morley's experiments evoked more questions that I would later pursue, but I was sure of one thing: light traveled at an extremely high rate of speed with one constant velocity which we have been able to measure to a high degree of accuracy. Other than that, I chose *not* to accept certain conclusions promoted as the "normalcy of relativity." I believed there were other explanations to be had and persisted throughout the years to search for such.

Initially, I postulated that it was the objects themselves (Earth, Moon, Sun, etc.) that were traveling at the velocity of light c, and that light gets its velocity by virtue of its emanation from these objects. The objects themselves, then, are in fact the *medium* with which light travels. For just as the trillions and trillions of molecules of air we breathe—each vibrating back and forth, tracing out a path of approximately 1,100 feet per second—are the medium which cause sound waves to travel 1,100 feet per second; so too are the trillions and trillions of astronomical bodies throughout the universe (such as the Earth, Sun, and Moon) the medium for the transmission of light at *their* speed through space of 186,000 miles per second. This, I believe, is why nothing can ever go faster than light. It is because physical bodies themselves must first propagate before their light—otherwise their images of photons—can propagate from them. Throughout my search for answers, it was my goal to either prove or disprove this postulate by assuming its truth and observing whether it would explain other well established physical phenomena in our universe as well as those phenomena as yet unexplained.

I continued to focus exclusively on the hypothesis I initially formulated and also to add another—one that would be another significant progression in the final thesis presented in this book.

The second hypothesis was a corollary of the first. If objects themselves are moving at the speed of light, then all objects must have *accelerated* from time-zero to their current velocity. This acceleration, taking into account what we know about other accepted physical parameters, should be calculable from what we already know about movement of bodies in our own solar system and the expanding universe according to Hubble's Law.

My progression of thought allowed me to formulate what I call, "The Three Truths." Hubble's Law and its description of the rate of universal expansion was one of them. The other two I had known and accepted first. Including Hubble's Law, they were: (1) Einstein's energy equation $E = mc^2$, (2) the Constancy of Light's Velocity, and (3) the Universal Expansion according to Hubble's Law. Accepting and understanding these three truths allowed me to build upon them to formulate what I felt was a fundamental unifying principle that explained their relationship with one another. This unifying principle is 2.8 Angstroms. Actually, it is more precisely stated as 2.8 Angstroms per second per second—an *acceleration*.

This value, 2.8 Angstroms per second per second, came after I sought to determine an acceleration necessary for objects in our solar system to be moving at the present velocity we attribute to light. It is equivalent to 2.8×10^{-10} meters, which is a minuscule length approximately equal to the diameter of an air molecule. I believed this value would have had to be consistent with other known and accepted physical parameters. At that point, I had little exposure to Newton's Second Law (F=ma), and to his universal law of gravitation ($F = GM_1M_2/r^2$), but took what might be called an "intuitive leap" in using the fundamental principle of the second law with some modification. I decided that if we were in fact moving through space at the speed of light, we would most certainly have to be accelerating in order to have gotten to that speed—or any other speed! To determine such rate of acceleration I simply took the force of gravity F_g at the surface of the Earth and divided it by the product of the Earth's mean density and mean radius times one-square meter. The resultant physical parameter of those key parameters is the *mass* of a *one-square meter column* m_c extending from the Earth's surface to its center. The use of the Earth's *unit column* for determination of *gravitational mass* rather than the Earth's *total* mass, appeared to be compatible with what mathematically contributed directly to the force of gravity at the surface of a body. The value I obtained from this calculation was 2.8×10^{-10} Newton per kilogram which, otherwise, is an acceleration of 2.8 Angstroms per second per second—a direct constant of proportionality between a body's magnitude of its force of gravity F_g at the surface and its mean column-mass m_c; again, such column-mass being the product of a body's mean *density ρ* mean *radius* r times *one-square-meter,* $1m^2$.

Of course, to have an acceleration a force is first necessary to cause the acceleration. The equation, $a_u = F_o/m_c$ does in fact reveal this. Likewise, it reveals that the magnitude of the universal Outside Force F_o pushing the Earth through space is 2.8×10^{-10} Newton *times* each kilogram derived from the mean mass of a one-square-meter column the length of the Earth's radius. Thus, the reason why all objects at the Earth's surface weigh 9.8 Newtons per kilogram (or accelerates in free fall near its surface at 9.8 meters per second per second) is simply because a constant Outside Force F_o of 9.8 Newtons is apparently acting upon, or otherwise pushing the Earth through space. This in turn causes the Earth to eternally accelerate at the constant rate of 2.8 Angstroms per second per second. In this same manner, the magnitude of the Outside Force acting upon, or pushing the Moon thru space is 1.6 Newtons, causing it to also accelerate at the rate of 2.8 Angstroms per second per second in direct relation to its column-mass of 5.8×10^9 kg. This then is the reason why *all* objects in the universe constantly accelerate at the eternal rate of 2.8 Angstroms per second per second.

How did this number fit in with "The Three Truths" I had previously accepted? From a cosmological perspective I saw that it fit in quite well! Not surprisingly, it turns out that Hubble's Constant H_o, (which denotes the rate at which the universe is expanding) appears to be one and the same value as the Universal Constant of Acceleration a_u. Why is this? Well, if the present estimated value for H_o is cut in half, $(H_o/2)$ as would be required to compensate for an "accelerating universe", and thereafter reduced to reflect only *one light second* instead of one million light years, H_o's value turns out then to be none other than a_u's value—both representing the same strict ratio between a body's force of gravity at its surface F_g, and its column-mass m_c.

If that is the case, then H_o's value can be refined even more so simply by determining a_u to a higher degree of accuracy. Remember too, the numerical value of G times $4\pi/3$ is a_u's value. How much closer can we wish to get to determine the real value of H_o? For whatever G's value is eventually refined to be in the lab, H_o, F_o and a_u will be precisely $4\pi/3$ times G.

When 2.8 Angstroms per second per second was divided into the current value for the speed of light, I obtained a distance traveled for the Milky Way from $t = 0$ of

17 billion light years, and an age for the particles of our galaxy to be about 34 billion years. This is about twice the currently accepted age as is classically determined by *light* years: 17 billion. However, this accepted value of 17 billion years assumes a *standard* light year for *all* time based on the *current* value of c. If however, c is in fact determined by the speed of the bodies themselves, then a more accurate value to use for c would be its *average* of 93,141 miles per second over the entire period of the expansion. And since a galaxy's own light year is used, its age in years will always be twice its distance traveled in light years from t = 0. This, is explained in detail in chapter 8 reconciling *my* estimated age of 34 billion years for our part of the universe and the commonly accepted age of 17 billion years.

This acceleration, 2.8 Angstroms per second per second, results, as previously stated, by dividing the force of gravity F_g at the surface of any body by its mean *column-mass* m_c. The ratio between this particular force and mass (as I have defined it independent of $a_u = F/m$) is always 2.8×10^{-10} Newton per kilogram, otherwise 2.8 Angstroms per second squared. I refer to this as the *Universal Constant of Acceleration* a_u, whose equation is simply: $a_u = F_g/m_c$ where a_u denotes the Universal Constant of Acceleration; F_g is the magnitude of the force of gravity at any point from a body's center of mass; and m_c is a body's mean column-mass, i.e., a body's mean density times its mean radius times one square meter. This derivation of acceleration using a *column-mass* rather than the total Newtonian mass was the turning point in my work. It clearly led to the *redefinition* of Gravitational Mass to be only the mass of a spherical body's *column*.

Assuming bodies themselves move at the speed of light provides another interesting result related to truth number one of The Three Truths I had earlier spoke of. Einstein's equation, $E = mc^2$, has been proved true. But why is it true? If all bodies in our solar system (all the particles it is comprised of) have accelerated at a constant rate of 2.8 Angstroms per second per second to a distance from $t = 0$ of 17 billion light years over a period of time of 34 billion years to their present velocity of 186,282 miles per second, then the Earth, for instance, is moving in space at that speed. At that speed the Earth really does possess a kinetic energy of mv^2. Thus, $E = mc^2$ is equivalent to $E = mv^2$ where v is equal to the Earth's current velocity—the speed we attribute to light. This provided a nonrelativistic explanation of why Einstein's famous energy equation, $E = mc^2$, worked. I believed it also provided further support for the hypothesis that the Earth is moving at the speed of light, and that light's propagation is secondary to our own basic propagation through space.

Realizing these things, it became reasonable to believe that the Earth was indeed accelerating at a rate of 2.8 Angstroms per second per second, and that gravity must have some relationship to this acceleration. Since I had used the force of gravity at the surface of the Earth to originally derive the universal acceleration, I was, in a sense, already assuming gravity and acceleration were closely related. I wanted to approach it however from a different angle to test my earlier methodology. Since gravity was a force, I concluded that it may be this intrinsic acceleration of all bodies that creates a body's force of gravity at its surface. This was in accord with Einstein's "Principle of Equivalence" which equated the effect on an object at rest in a gravitational field with the effect on the same object inside a rocket ship being accelerated upward in a gravity-free environment. Was this true because gravity was—in reality—caused by an acceleration?

Newton's law of universal gravitation would naturally come to occupy much of my time. I was not entirely comfortable with this law since G, his critically necessary universal constant in his gravitation equation, did not, in and of itself, have a direct *linear* relationship to the magnitude of a body's surface gravity as did the constant a_u. I also came to realize that G's value, whose quantity must normally be determined in the laboratory, was given physical units which had no explicit meaning. Its units: $m^3kg^{-1}s^{-2}$ and Nm^2kg^{-2} are said to be *peculiar* to G, but were necessarily used as such in order to be consistent with Newton's Law of Gravitation. The units for G were provided by Newton to balance his equation but they said little about what caused gravity. These "peculiar" units made me scrutinize Newton's law of gravitation even more. I began to question if there could have been a better, simpler, equation to describe gravity. This provided further impetus to determine if gravity could be shown to be related to acceleration, and if the equations $a_u = F_o/m_c$ and $a_u = F_g/m_c$ were in fact valid.

I found that Newton's second law could, in fact, be used to determine the gravitational force of attraction at the surface of any body if the Universal Constant of Acceleration (2.8 Angstroms per second per second) were used in the equation with the mean *column-mass* of the body. If 2.8 Angstroms per second per second is accepted, then it can be seen that a simple *linear* equation may be used to determine the force of gravity at a body's surface, i.e., its gravitational field strength, rather than the currently accepted non-linear equation using G. This simpler equation, utilizing a_u's value instead of G's value, worked because bodies such as the Earth and other planets and stars, are apparently being *accelerated* through space as a result of a *force* on that body. Better still, some Outside Force F_o of a universal nature is eternally acting, or otherwise pushing on the Earth and all other bodies, resulting in a constant acceleration. However it might be viewed, the resulting quantities are described by Newton's second law, $F = ma$ with the modification to $F_g = m_c a_u$.

The evidence clearly indicates that some outside source is driving each and every body, particle, etc., with a force of 2.8×10^{-10} Newton per kilogram of the body's mean column-mass. So that there is a *constant* Outside Force, for example, of 9.8 Newtons pushing the Earth through space; or 1.6 Newtons pushing the Moon through space, or approximately 10^{-7} Newton pushing on all protons, which results in *all* bodies accelerating at the same constant rate of 2.8 Angstroms per second per second. This is the case whether we are speaking of planets, protons, or electrons. Every body, every particle is apparently being driven by a force of 2.8×10^{-10} Newton for each kilogram of mass—hypothetical or real—calculated from a particular body's mean column-mass; so that the net resultant acceleration is the same for all bodies— 2.8 Angstroms per second per second. It is this precise rate of acceleration that creates their gravitational attractive forces. These attractive forces, in turn, are directly proportional to the body's column-mass m_c, otherwise, a body's mean radius times one-square meter times its mean density. Therefore, the total Outside Force F_o pushing the body through space—causing them all to accelerate at the same constant rate of 2.8 Angstroms per second per second—s of the same precise magnitude as is the force of gravity F_g per kilogram of mass at the body's surface.

This book, THE NEW GRAVITY presents a detailed discussion of my first book, 2.8 ANGSTROMS, but with much emphasis on gravity, per se. It details much about how an ongoing acceleration might explain the wave-particle dualism and the momentum-position dilemma of the electron; plus other possible explanations for such items as quasars, and the 3-degree microwave background radiation. Most importantly, it deals with what has occupied physicists minds for decades: a Unified Theory for Gravity and Electromagnetism.

1

The New Gravity

Einstein's Theories of Relativity set physics reeling on a whole new course, one which radically changed the scientific community's concept of Space and Time. His contributions to science included, amongst other things, his famous equation on Mass and Energy $E = mc^2$, which is still today considered the most famous equation of all time.

I believe however, that in spite of the many good and positive works Einstein contributed, he may have "unwittingly" steered the scientific community slightly off course when he made the far reaching assumption that *the speed of light is an absolute and unchanging constant of nature throughout all space and time.* He later raised his assumption to a *postulate*, and thus, it became a major turning point in the way we think about physics.

In this book it will be advocated quite to the contrary, that the speed of light is NOT an absolute and unchanging constant of nature, but that there appears to exist an

"Outside Force" of 2.8×10^{-10} Newton acting upon each Column-Mass kilogram throughout all space and time which does in fact cause light's velocity to *continuously increase with time*. This Outside Force F_o apparently creates a Universal Constant of Acceleration a_u of approximately 2.8×10^{-10} meter (2.8 Angstroms) per second per second which is imposed upon all material bodies in the universe—from electrons and protons to planets and stars.

Consequently, the ensuing and *varied* velocities of all material bodies throughout the universe become the vehicle, the means in which Light and all other forms of Electromagnetic phenomena derive their energy of propagation. In reality then, it is the material bodies of the universe (the protons, neutrons, and electrons, etc., comprising the galaxies of stars and planets) which actually possess the *real* and *basic velocity*—the energy which the velocity of light itself derives from—increasing as they do by a mere 2.8 Angstroms for each second of time. So that in 34 billion years for example, the materials of our own galaxy have accelerated at this constant rate to their present velocity of 186,282 miles per second while having traversed a distance of 17 billion of our present light years in one general direction from time-zero, $t = 0$, where it is generally believed the center of mass of our universe originated.

Therefore, it will be shown that the Universal Constant of Acceleration a_u is always equal to F_o/m_c, where F_o is the total Outside Force in Newtons pushing the body thru space, and m_c denotes a body's *Column-Mass*, a new physical parameter which is the product of a body's mean *density* ρ and mean *radius* r times *one square meter* $(1m^2)$, which otherwise is a new definition for *gravitational* mass. So that even for a proton which is said to have a radius of 10^{-15} meter, its hypothetical column volume would be 10^{-15} *cubic* meter, but only however for purposes of determining a_u and the magnitude of F_o at its surface. For example, in using the physical parameters of the proton, we determine a_u's universal value by using the *new column-mass* equation, $m_c = \rho r \cdot 1 m^2$, thus

$$a_u = F_o / m_c \qquad\qquad (1-1)$$
$$a_u = 1.2 \times 10^{-7} N \ / \ 4.3 \times 10^2 kg$$
$$a_u = 2.8 \times 10^{-10} ms^{-2}.$$

This equation $(a_u = F_o / m_c)$ maintains a constant ratio between a body's Outside Force F_o and its column-mass m_c. This ratio, $2.8 \times 10^{-10} ms^{-2}$ is the Universal Constant of Acceleration a_u which actually creates a body's force of gravity F_g at its surface.

It follows then that the magnitude of the *force of gravity* F_g at any body's surface would be $F = m_c a_u$. Using the physical parameters of the Earth for example, the force of gravity F_g at its surface would simply be

$$F_g = m_c a_u \qquad\qquad (1-2)$$
$$F_g = (3.5 \times 10^{10} kg) \ (2.8 \times 10^{-10} ms^{-2})$$
$$F_g = 9.8 N.$$

That force F_g, of 9.8 Newtons acting upon, or otherwise, *pulling* at each kilogram of mass resting upon the Earth's surface is due to the same equivalent *Outside Force* F_o constantly acting upon or *pushing*, the Earth through space, so that both equations are precisely the same, thus, the total Outside Force F_o is the total force of gravity F_g at the surface of all bodies. Once again, in the case of the Earth, the Outside Force F_o is

$$F_o = m_c a_u \qquad\qquad (1-2a)$$
$$F_o = (3.5 \times 10^{10} kg) \ (2.8 \times 10^{-10} ms^{-2})$$
$$F_o = 9.8 \ N.$$

The numerical value, 2.8×10^{-10}, is derived as the ratio which is always found to exist between a body's Outside Force F_o and the product of its mean density ρ and mean radius r. This value, 2.8×10^{-10}, will also be found to be the *product* of Newton's Universal Constant of Gravitation G and $4\pi/3$, such that

$$a_u = G(4\pi/3) \qquad\qquad (1-3)$$
$$a_u = (6.672 \times 10^{-11}Nm^2kg^{-2})(4.189)$$
$$a_u = 2.795 \times 10^{-10}Nm^2kg^{-2}.$$

When G's units are shown as $m^3kg^{-1}sec^{-2}$, then of course a_u's units will likewise be equivalent. However, what will more accurately be shown for a_u's units, and indeed its precise value, is that its units will instead reflect an *acceleration* (ms^{-2}) and its value, 2.795×10^{-10}, will be seen to be a Direct Constant of Proportionality between Force F and mass m just as with Newton's second law: $F = ma$!

It will also be shown that the product of G and $4\pi/3$ is not simply all there is to a_u's value. A much more interesting connection between ($4\pi/3$'s) 4.189 value and a_u's 2.795 Angstrom units will be revealed in detail later in the chapter on how a_u's value directly derives (in the most fundamental sense) from the *basic* one-kilogram spherical mass whose physical parameters and their precise values *all* stem from $4\pi/3$'s value of 4.189. To summarize, it turns out that the acceleration due to gravity at the surface of this basic one-kilogram sphere is precisely that of a_u's value, $2.795 \times 10^{-10}ms^{-2}$! Again, there will be much more detail on this most interesting fact later in the chapter.

To return however to equation 1–3 above, it will be found that the physical units of G (Nm^2kg^{-2} and or $m^3kg^{-1}sec^{-2}$) are actually *peculiar* to G but were *chosen* by necessity in order to make Newton's Law of Universal Gravitation, $g = GM/r^2$ and $F = GM_1M_2/r^2$, *dimensionally correct.* It will be found that, in fact, G's units apply *naturally* to a_u's value of 2.795×10^{-10}, and this item too will be elaborated upon later in the chapter.

Although G's lengthy units (Nm^2kg^{-2} or $m^3kg^{-1}sec^{-2}$) do in fact apply *naturally* to a_u's value of 2.8×10^{-10} (a_u's value will at times be shown as such instead of $2.795 \times 10^{-10}ms^{-2}$) a_u will instead be shown to have associated with its value the physical unit of ms^{-2}—*strictly an acceleration*—as it properly and naturally should be.

The reason for *that* just stated is simply because a_u is a *direct constant* of *proportionality* between the *Force* of gravity F_g at a body's surface (or, equivalently, the Outside Force F_o acting on it) and the body's *column-mass* m_c. Such column-mass—a new physical parameter *redefining* gravitational mass—is the product of a body's mean density ρ and mean radius r times one-square meter $1m^2$; so that, using for example, the physical parameters of the Earth, we would derive for a_u's value

$$a_u = F_o/m_c \qquad\qquad (1-4)$$
$$a_u = 9.8N/3.5 \times 10^{10}kg$$
$$a_u = 2.8 \times 10^{-10}ms^{-2}.$$

As can be seen, equation 1–4 is actually Newton's second law. However, in this context it shows that each and every body in question (Earth, Moon, Sun, proton, electron, etc.) is *eternally accelerating* thru space at a constant rate of 2.8×10^{-10} meter per second per second.

For an electron, that acceleration is a natural and eternally constant increase in its velocity for each second of time by a distance or length equal to about 67-million times its own point like diameter. This is most likely the reason why particle physicists have so much difficulty determining—*with precision*—the electrons *momentum* and *position* simultaneously. If they were to accept the premise of their natural and eternal acceleration, there would likely be no more Heisenberg Uncertainty principle to cope with. Because all electrons, protons, and neutrons are constantly and eternally accelerating, and therefore, so too are we here on Earth and Earth itself, then so too is the velocity of light increasing by this same amount (2.8 Angstrom units per second per second) since light must then be considered to be a secondary effect emanating as a result of the basic and or primary velocity of solid bodies thru space.

The whole argument to be made for the New Gravity is simply that Mass is herein being *redefined,* in that, whereas a body's *Inertial* and *Gravitational* mass have always been classically represented to mean their TOTAL mass (derived at with the formulation of $4\pi/3 \cdot r \cdot r \cdot r \cdot \rho$); we instead proclaim a body's Gravitational Mass to be *only* the body's *Column-Mass* m_c, which otherwise is the product of the body's mean *density* (ρ) times its mean *radius* (r) times *one-square-meter* ($1m^2$). In other words, the formulation for the *new* Gravitational Column-Mass m_c is

$$m_c = \rho r \cdot 1 m^2 \qquad (1-5)$$

while one is reminded that $4\pi/3 \cdot r \cdot r \cdot r \cdot \rho$ (which represents a body's *total* mass) is the classical formulation for Inertial Mass of spherical bodies.

That then is the *redefinition of gravitational mass* which the theory in this book will adhere to, i.e., Column Mass for Gravitational Mass ($\rho r \cdot 1 m^2$); and Total Mass for Inertial Mass ($4\pi/3 \cdot r \cdot r \cdot r \cdot \rho$). That will be the distinction between the two different masses, in which it is conclusively shown they are *not* equivalent. This of course contradicts classical theory in which both types of mass have always been believed to be proportional to one another. Now however, one can utilize the Formulation of Newton's Second Law ($F = ma$) in *both* cases, but mass must now be defined in a different manner between the two.

Getting off the subject of mass and its definition for awhile, let's go back to Newton's G in order to understand a little more about it. It is a difficult task to find any elaboration at all about "G" throughout the literature. The minuscule value (6.672×10^{-11}N) for Newton's Universal Constant of Gravitation G was basically and or fundamentally established to be equal to the mutual but opposing forces of attraction between two one-kilogram bodies, each being *not more* than one-meter in diameter, whose centers of mass are *one-meter apart*. This is Newton's Law of Universal gravitation, $F = GM_1M_2/r^2$, in which M_1 and M_2 are each one-kilogram whose centers of mass are one-meter apart. They can also be thought of as being point masses, as Newton's law teaches, or any size up to one-meter in diameter in order for one-single unit of force of Newton's G (6.672×10^{-11}N) to apply between them, as indeed it does. These were the basic masses and distance utilized to give G its constant value.

If however, we now consider only a *single* one-kilogram mass whose radius is one-meter (two-meter diameter) or *smaller*, Newton's law is then classically applied as: $g = GM/r^2$. This, substantially, is what Newton's law is all about. If we stick to this more *simple* concept, we find out what we're really after; and that is: at any distance from M's center of mass (whether at its surface or any point in space) what is the magnitude of its gravitational field strength, g? That is really *all* one needs to determine— what is g's magnitude of gravitational intensity from the center of any body's mass out to the center of mass of any other object in question? Using for example the Earth's mass of approximately 6.0×10^{24}kg, if we wish to know g's value for the acceleration of the Moon towards the Earth due to the Earth's gravitational intensity at the

Moon's position in space, we would simply multiply G's $6.672 \times 10^{-11} \text{m}^3\text{kg}^{-1}\text{sec}^{-2}$ times the Earth's mass of $6.0 \times 10^{24}\text{kg}$ and divide their product by the square of the distance (r^2) between their centers of mass $(3.844 \times 10^8\text{m})^2$, and we have for the Earth's g (F_g with the New Gravity) at that distance, $2.7 \times 10^{-3}\text{ms}^{-2}$. This then will be the acceleration of *any* freely falling body at that precise distance from the Earth due to its force of gravity at that point in space—whether it be the Moon or an apple, each will be pulled towards the Earth with an acceleration of $2.7 \times 10^{-3}\text{ms}^{-2}$.

Let's suppose now that the apple were a *large* apple having a mass of exactly one-kilogram. The mutual but opposing gravitational force of attraction F between the Earth and the one kilogram large apple at that point in space would be $2.7 \times 10^{-3}\text{N}$ each. In other words, the Earth would pull at the apple—located at the point in space where the Moon is—and the apple would pull at the Earth from that far out, each with a force F of 2.7×10^{-3} Newton. In the same context, as the apple and the Earth, both the Moon and the Earth would pull at each other with a gravitational force of attraction of $2.7 \times 10^{-3}\text{Nkg}^{-1}$ times the Moon's total mass of $7.35 \times 10^{22}\text{kg}$ for a total force of attraction of 1.9×10^{20} Newtons each. Note that to determine mutual but opposing gravitational forces, only one of the two body's physical dimensions (either one) need be considered.

If we do the calculation for determining how fast the Earth is *accelerating* towards the Moon, we will find that to be $3.32 \times 10^{-5}\text{ms}^{-2}$. This tells us that the Moon's acceleration of $2.7 \times 10^{-3}\text{ms}^{-2}$ towards the Earth is actually 81 times greater than the Earth's acceleration towards the Moon. It also tells us that this is so simply because the Earth has approximately 81 times the mass of the Moon. This is why Newton's third law functions as it does. There is always in the math an equal but opposite force between *any* two bodies simply because we *reverse* their total masses with their actual field strengths, i.e., the ratio of their *masses* (Earth to Moon) are 81-to-1, and the ratio of their *field strengths* at each other's center of mass (Moon to Earth) is 1-to-81. Reversing two sets of values, as seen here, simply neutralizes the picture, and so, Newton's Third Law: *"For every action there is an equal but opposite reaction."*

Look at the 1.0 kg apple and the 6.0×10^{24}kg Earth. The Earth pulls at the apple causing the apple to accelerate towards it at a rate of 2.7×10^{-3}ms^{-2}. But the 1.0 kg apple pulls at the Earth causing the Earth to accelerate towards it (the apple) at a rate of 4.5×10^{-28}ms^{-2}. These acceleration rates have a ratio of 6.0×10^{24}-to-1. The ratio of the apple's total mass to the Earth's total mass, however, is an equal but opposite ratio of 1-to-6.0×10^{24}. So there we have once again the simplistic manner in which the *third law* operates. That is how Newton's law of universal gravitation functions—why one body always pulls at a second body with an *equal* but *opposite* force.

If we never utilize Newton's $F = GM_1M_2/r^2$ it would be no catastrophe, because if we utilize Newton's simpler version, $g = GM/r^2$, we only need multiply g's value times *any* particular mass located at the point where we determined a value for g. One last item to stress is that both, g's acceleration rate, and g's field strength (classically) in Newtons per kilogram each have the same numerical value at all times. For example, the Earth's rate of acceleration for *any* freely falling body near its surface due to Earth's gravity, is 9.8 meters per second per second. This means, for example, that both a thousand kilogram sphere of gold and a one kilogram sphere of styrofoam would each hit the Earth's surface at the same moment if dropped from 1,000 feet at the same precise moment; that is, if the one-kilogram styrofoam sphere were not slowed down due to friction with the air. However, when they both finally are resting on the Earth's surface, the Earth's force of gravity of 9.8 Newtons at that point would pull at the gold sphere with a force of 9.8 Newtons for each kg of its total mass for a total force of 9,800 N; and it would pull at the styrofoam sphere with a total force of only 9.8 N.

In order to determine the precise value for Newton's constant of universal gravitation G, it is necessary to find this value experimentally in the lab. However, Newton was able to estimate G's value by determining a good approximation for the Earth's mean density and using his formulation as here shown

$$G = gr^2/M. \tag{1-6}$$

Of course g was known for the Earth's surface, and the Earth's radius was also known at the time. So Newton, who estimated the Earth's density to be 5.5 times that of water's, 1,000 kg/m^3, was able to derive a close approximation for G of $6.7 \times 10^{-11} \text{m}^3 \text{kg}^{-1} \text{sec}^{-2}$. Today's best value is $6.672 \times 10^{-11} \text{Nm}^2 \text{kg}^2$.

Technically however, equation (1–6) entails doing the following derivations for determining G's value when both the radius and density are known along with g. It is emphasized here that a body's *total* mass cannot necessarily be taken for granted, thus, in essence

$$G = g \cdot r \cdot r / (4\pi/3) r \cdot r \cdot r \cdot \rho \tag{1-6a}$$
$$G = 6.672 \times 10^{-11} \text{Nm}^2 \text{kg}^{-2}.$$

Equation(1–6a) results in G's value being $6.672 \times 10^{-11} \text{Nm}^2 \text{kg}^{-2}$ no matter which body the physical parameters are taken from. The units would come out as $\text{m}^3 \text{kg}^{-1} \text{sec}^{-2}$ if g's units were expressed in ms^{-2} instead of Nkg^{-1} as is implied from (1–6a).

The question now is: why should a_u's value (2.8×10^{-10}) be more useful than G's value? The answer is that since a_u's value is a strict constant of proportionality between *force* and *mass* (the new column-mass), this then indicates that all physical bodies throughout the universe (protons, electrons, planets, stars, etc.) are eternally *accelerating*—as equation (1–4) indicates—at a constant rate of $2.8 \times 10^{-10} \text{ms}^{-2}$ due to some Outside Force of that same equivalent numerical quantity of 2.8×10^{-10} Newton per kilogram. However, this Outside Force F_o is now 2.8×10^{-10} Newton for each kilogram calculated from a body's *column*-mass. Thus, if for example we utilize the Earth's physical parameters, we would have for the *total* Outside Force F_o acting upon the Earth

$$F_o = m_c a_u \qquad (1-7)$$
$$F_o = (3.5 \times 10^{10} \text{kg})(2.8 \times 10^{-10} \text{ms}^{-2})$$
$$F_o = 9.8 \text{ N}$$

and this, precisely, is why any other body resting on the Earth's surface is weighted down with a force of 9.8 Newtons for each kilogram of its total mass.

Contrast deriving the value for the New Gravity's Universal Constant of Acceleration a_u in equations (1–8) and (1–8a), with Newton's classical formulation of equations (1–6) and (1–6a), where for a_u we simply have a *linear* formulation with less hypotheses, such that

$$a_u = F_o / m_c. \qquad (1-8)$$

Compare now the new formulation of (1–8a) below with Newton's formulation in (1–6a) in which the equations apply to any and all bodies, thus we have for a_u

$$a_u = F_o/m_c \qquad (1\text{--}8a)$$
$$a_u = F_o/\rho r \cdot 1 m^2$$
$$a_u = 2.8 \times 10^{-10} ms^{-2}.$$

Once again, in making a case for the New Gravity, it is simply that I have redefined *gravitational mass* in the context *solely* as a body's *column-mass* m_c, which otherwise is the product of the body's mean (average) density and mean (average) radius times one-square-meter, and that is that! It all follows, most accommodatingly, Newton's Second Law of Motion, $F = ma$. Thus, it strongly indicates that there is good and sufficient reason to believe that both, *gravity* and *acceleration* are—in reality—equivalent after all is said and done. It is, in other words, the acceleration of all bodies thru space at an eternally constant rate of $2.8 \times 10^{-10} ms^{-2}$ that is creating the forces of gravity F_g at both, their real and their hypothetical surfaces.

This acceleration itself appears to be a direct result of some as yet unidentified Outside Force F_o of an equivalent 2.8×10^{-10} Newton acting upon each body with a total force in strict proportion to a particular body's column-mass m_c, and NOT its *total* mass.

Naturally, it is a body's *total* mass that actually creates its attractive force; but it appears to be *only* its *column-mass*, i.e., its density times its radius that is directly related to its attractive forces.

What is different about the New Gravity is that one need only learn to use mass in a different context than what has classically been the practice. This is simply shown to be the specific mass—either *real* or *hypothetical*—of a *column* having a cross section of *one-square-meter* ($1m^2$) whose length is the body's mean radius (r) times the body's mean density (ρ).

For the special case of a *hypothetical* column, this is derived when, for example, one is determining the *column-*

mass for a proton whose radius is a minuscule 10^{-15} meter. To determine a proton's hypothetical column-mass, its length (radius) must hypothetically be given a cross-section of *one-square-meter* as the formula calls for. So now the proton's column-volume v_c, is hypothetically $10^{-15}m^3$. The proton's supposedly *real* total volume to begin with is only approximately $10^{-45}m^3$. One may ask then: "so how can it have a column-volume of the much greater value of $10^{-15}m^3$? This is thirty (30) powers of ten greater in volume than the proton's real total volume. The answer would have to be that this is apparently the way *gravitational mass* must indeed be determined; for once the MKS system (Meter, Kilogram and Second) and all related matter to *gravity* and *gravitation* had been established and the rules laid down, we should had—as I am attempting to do in this book—defined mass in its gravitational context somewhat different than inertial mass.

Inflated Bodies Replace Newton's Point Mass Bodies

What else does this New Gravity with its *redefined* gravitational mass reveal that is significantly different from Newtonian gravity?

Clearly, the New Gravity *reverses* Newton's *point mass* principle! It reveals that instead of hypothetically DEFLATING a body's total volume down to a point particle and still retain its real total mass, we must now hypothetically INFLATE and *uniformly distribute* a body's total mass anytime the magnitude of its force of gravity F_g must be known for any point in space away from its real surface. I would emphasize here however, that *uniform distribution* of a body's total mass *always* applies, whether we are referring to a body's *real* volume, or to its hypothetically *inflated* volume. This is implied throughout the book with the nomenclature concerning a body's *density,* in that it is always stated as MEAN density—and this means a body's *overall mean average density*.

A good example of the above is that even if we have an extreme situation where 90% of a star's total mass is concentrated within the volume of the inner 10% of its radius, and only 10% of its total mass is spread through-out the volume of its outer 90% of its radius, the *mean density* for such a star would still be its *total* mass divided by its *total* volume. It would read *kilograms* per *cubic meter uniformly* distributed throughout the whole of its *real* or *hypothetically* inflated volume.

When we talk about a body's Column-Mass, what we are speaking of, in essence, is nothing more than its real or inflated radius (distance) times its real or diluted density out to any point from its center of mass where another body is located, and therefore attracted by it. That in fact follows closely what has always been the *fundamental axiom* of Newton's law of universal gravitation, which partially states: The Force between any two particles is *"an attraction acting along the **line** joining them."* Notice I emphasize *"**Line**,"* for is it not the product, *solely*, of a body's *density* and *radius*? Is it not a *line* joining two bodies from each other's centers of mass? And is not this the line—the Column-Mass Line—which is the specific physical parameter that directly and or proportionately determines in a *linear* manner, the magnitude of a body's gravitational field strength, otherwise, its force of gravity F_g at any point from its center of mass.

If we go back then, again to the Earth's force of gravity F_g way out in space at the position of the Moon's center of mass, we would determine the Earth's inflated column-mass m_c by cubing the distance (r^3) between them, and multiplying that product times $4\pi/3$. If we next divide this much greater hypothetical volume into the Earth's real mass of 5.976×10^{24}kg, this results in an extremely diluted mean density of only 2.512×10^{-2}kg/m³;

which when multiplied times the Earth's distance (r) to the Moon, will give us the extended column-mass (m_c) from the Earth's center of mass out to the Moon's center of mass.

Thus, in calculating for the Earth's hypothetical column-mass m_c out to the Moon's center of mass, we would have

$$m_c = \rho r \cdot 1 m^2 \qquad (1-9)$$
$$m_c = (2.512 \times 10^{-2} kg/m^3)(3.844 \times 10^8 m)(1 m^2)$$
$$m_c = 9.656 \times 10^6 kg.$$

To determine next the magnitude of the force of gravity F_g which the Earth exerts out at the Moon's position in space (remember, the Earth is now hypothetically *inflated* with a radius out to the Moon's center of mass, 384,400 kilometers away) we have in place of Newton's classical formulation ($g = GM/r^2$) the New Gravity's Column-Mass formulation of $F_g = m_c a_u$; so that the force of gravity upon each kilogram of mass at the Moon's distance, or *any* other body equal to the Moon's distance from the Earth's center of mass, will thus be

$$F_g = m_c a_u \qquad (1-10)$$
$$F_g = (9.656 \times 10^6 kg)(2.8 \times 10^{-10} ms^{-2})$$
$$F_g = 2.7 \times 10^{-3} N.$$

This is the magnitude of the gravitational field strength, 2.7×10^{-3} Newton, with which the Earth pulls at *each* kilogram of mass of *any* body that is located in any direction from the Earth's surface out to a distance of 384,400 kilometers. Of course, to determine the *total* force (F) or *weight* (w) of the Moon if it were hypothetically resting on the Earth's hypothetically inflated surface—which is now blown up out to the Moon's center of mass—we would multiply the Moon's total mass of 7.35×10^{22}kg times the Earth's force of gravity F_g at that point, which of course is extremely reduced to 2.7×10^{-3} Newton, so that the Moon would weigh if it were indeed sitting upon the Earth's hypothetical surface

$$W_M = M_M F_g \qquad\qquad (1-11)$$
$$W_M = (7.35 \times 10^{22}\text{kg})(2.7 \times 10^{-3}\text{Nkg}^{-1})$$
$$W_M = 1.98 \times 10^{20}\text{N}.$$

Newton's Third Law

The value, 1.98×10^{20}N in equation $(1-11)$ is also the total force with which the Moon pulls at the Earth, or what the Earth would weight if *it* were resting on a hypothetically inflated Moon which is blown up all the way out to the Earth's center of mass 384,400 km away. These are equal but opposite forces which obey Newton's third law of motion which states: "Whenever one body exerts a force on a second body, the second body exerts an equal but opposing force on the first body." Again, I speak here about how simple this law is to understand. It arises from the fact that we have a set of opposite values of the two body's total masses of 81-to-1 and gravitational field strength pulling at them of 1-to-81, which alone is all one needs to visualize to understand the concept of the third law's *equal* but *opposite* force axiom.

We have then for the Moon's total force of attraction for the Earth, an equal but opposite force, just as the Earth has for the Moon in equation $(1-11)$, thus

$$W_E = M_E F_g \qquad\qquad (1-12)$$
$$W_E = (5.976 \times 10^{24}\text{kg})(3.32 \times 10^{-5}\text{Nkg}^{-1})$$
$$W_E = 1.98 \times 10^{20}\text{N}.$$

As is readily seen, the Earth's total mass is 81 times the Moon's total mass, but then the gravitational intensity created by the Moon at the Earth's location is only 1/81th of the intensity which the Earth's field creates at the Moon's position. As can be seen, one parameter's value cancels out the other, and the playing field appears to be level. But, in reality, let's face it, the Earth's gravitational field strength is causing the Moon to *fall* towards it, i.e., to accelerate towards it, at a rate 81 times faster $(2.7 \times 10^{-3} \text{ms}^{-2})$ than the Moon causes the Earth to accelerate towards it $(3.32 \times 10^{-5} \text{ms}^{-2})$.

What those pulls on each other actually amount to is that the Moon falls towards the Earth at a rate (in British units) of slightly more than 1/10th of an inch per second per second. For the Earth, it is supposedly falling towards the Moon at a rate of slightly more than 1/1,000th of an inch per second per second. However, because of the Moon's orbiting velocity around the Earth, they do not really advance towards each other, and therefore, they maintain their original orbits. In fact, for other reasons, they are believed to be drifting away from each other by two or three centimeters per year—this is about *one-inch* in British units.

In any case, when one considers that the Moon's gravitational intensity here at our position—384,400 kilometers away from it—is just strong enough to cause the Earth to fall towards it at the minuscule rate of only about 1/1,000th of an inch per second per second, one might question how such a minute pull by the Moon upon the Earth could cause such large displacements of water, as is the case with the Earth's ocean tides. It might be more plausible instead to wonder if some part of the tidal action can be due to the *constant acceleration* of the Earth thru space at the rate of $2.8 \times 10^{-10} \text{ms}^{-2}$. In other words, as each point of the global Earth rotates to the position of its forward thrust vector of its continuing journey through space in the opposite direction away from time-zero, $t = 0$, that is the moment when the Earth's constant acceleration thru space causes the waters at that point to be displaced. Think about it as when an automobile you are riding in *accelerates*. Your body tends to be displaced in the opposite direction, pressing against the back of the seat. Now if that were a bucket of *water* instead of your body on the seat, the water will actually *tide up*.

Of course, the strong case for the Earth's ocean tides is classically accepted as being due to the Moon's gravitational pull on the Earth, albeit only 1/1,000th of an inch per second per second. It is however, still something to think about if in fact the Earth, Moon, Sun and galaxies throughout the universe are all actually *accelerating* thru space as is postulated throughout this book. For as has been shown, it is precisely the acceleration of the Earth thru space (allegedly caused by some Outside Force) that creates the attractive force which we experience and refer to as gravity at its surface.

Problems in Applying Physical Units

In developing the basic equation to explain the New Gravity there is one problem which has been difficult to overcome. It relates to the physical units for the force of gravity F_g. This is the symbol which denotes what "g" in Newton's equation for gravity denotes. Where Newton's "g" classically has units of Nkg^{-1} or ms^{-2}, F_g in the context of the New Gravity has the unit only of "N" for denoting a Newton of Force, and not Nkg^{-1}. This unit, "N" for Newton of Force is justified however as being all that is necessary. (Note that we are always working with the SI units and no other system unless it is spelled out as such.)

The question was this: In developing the units for F_g, must I have to define the resultant unit of the equation, $F_g = m_c a_u$, in Newtons per kilogram? I've tackled it somewhat like this: Let us say that at the Earth's surface there are *no* other bodies to be attracted. There still exists however an attraction, a force of gravity of 9.8 Newtons at the Earth's surface which pulls with the 9.8 Newton force at all other bodies, objects, etc., resting at its surface. Whether that second body has a mass of one-kilogram or one-thousand kilograms, the force is still the same upon each separate mass. However, each rests upon the surface with a pressure or *weight* equal to the product of the 9.8 Newton force and their *total* mass. In this respect, it is indeed Nkg^{-1}; but in the sense of describing the actual intensity of the force, it does not appear to be wrong to describe that particular force, at say, point A or Point B, as being 9.8 Newtons. Whether it is per this or per that, we are still strictly referring to the *magnitude* of the force in "Newtons." We needn't put *labels* on everything and can, if we wish, do without them as long as we make *clear* what we are saying. The force of 9.8 Newtons therefore is just that—a force of 9.8 Newtons at the Earth's surface. It is, in fact, the *equal* but *opposing* force to the *total Outside Force* F_o of 9.8 Newtons acting upon, or pushing the Earth thru space. This, in fact, is an ideal example of Newton's third law: for every action (F_o) there is an equal but opposite reaction (F_g)!

In free fall near the Earth's surface, all bodies equally fall at $9.8 ms^{-2}$ whether they have a mass of one-kilogram or one-thousand kilograms. In any case, however that point may be argued against F_g having dimensions of *Newtons only*, and *not* Newtons *per kilogram*, F_g's unit is declared to be what it is, "N", at least in the context of the New Gravity. After all, did not Newton have to *add* to G's natural value, $6.672 \times 10^{-11} N$, the *additional units* of

m^2kg^{-2} in order to make his Universal Law of Gravitation work? Cannot I enjoy just a little of what Newton did to keep everything in line. He added a square meter (m^2) and a square kilogram (kg^{-2}) to G's units in order for G to be consistent with his gravitational equation, $F = GM_1M_2/r^2$ and $g = GM/r^2$. I have neither added nor deleted any unit to my Constant of Universal Acceleration a_u, in order for it to be consistent with my gravitational equation, $F_g = m_c a_u$. In this equation however, where F_g is the equivalent of g in Newton's equation, F_g has only the unit of Newton's N, while g has the units of Newtons per kilogram Nkg^{-1}. However, the per *this* or per *that*, or per *kilogram*, is the *label* I mentioned earlier which has classically been attached for additional clarity to understand g's function. The *alternate* units of g are ms^{-2}; and the alternate units for F_g are $kgms^{-2}$. Here too there *appears* to be a problem with F_g, since we are looking for just an *acceleration*, ms^{-2}, as in Newton's g. However, $kgms^{-2}$ has classically been the equivalent of a Newton "N". In conclusion, there is in fact this problem which still must be resolved. I can only leave it to the experts to fix it, if indeed it must be fixed; although I can only hope that its *not broke* in the first place, and that too is for the experts to determine.

As for the New Gravity's Outside Force F_o, there is no problem with that particular unit as it does indeed strictly call for a force, for example, of 9.8 *Newtons* for the Earth, and 1.6 *Newtons* for the Moon and so on. This is the Outside Force (the fundamental universal force of $2.8 \times 10^{-10}N$ times each kilogram of a body's column mass) which *mysteriously* acts on every body in the universe. As stated previously, it exemplifies Newton's third law *ideally*, since for the Earth, there is an Outside Force $\vec{F}_o$ of 9.8 Newtons pushing on it, and there is an *equal* force $\vec{F}_g$ of 9.8 Newtons acting on it in the *opposite* direction! The total Outside Force acting upon or pushing the

Earth thru space is due to the Universal Outside Force $_u\vec{F}_o$ of 2.8×10^{-10} Newton acting on *each kilogram* kg of a body's column-mass m_c; and the total Force of gravity $\vec{F}_g$ at the Earth's surface is also due to the same identical factor. As a result, the Universal Outside Force of 2.8×10^{-10} Newton acting upon each kilogram of the body's column-mass, causes the body as a *whole* to attain a constant acceleration thru space of $2.8 \times 10^{-10} ms^{-2}$. Likewise, F_g's value (g's value in Newton's equation) is always equal to $2.8 \times 10^{-10} ms^{-2}$ at *any* body's surface for *each* and *every kilogram* we count in the body's Column-Mass m_c.

Some History of Newton's G

We go on now to some of the more basic mechanics and history of Newton's Universal Gravitational constant, G. The value for G is derived as the force between *two one-kilogram masses one-meter apart*. The force of attraction between these two one-kilogram spherical masses whose centers of mass are one-meter apart, is mutually, but opposingly, 6.672×10^{-11} Newton. As stated before, the units, Nm^2kg^{-2} were chosen as such in order that G's value would be compatible with Newton's Law of Universal Gravitation, $F = GM_1M_2/r^2$. The units, $m^3kg^{-1}sec^{-2}$ were also chosen units in order that G's value would be compatible with Newton's *simpler* form of his gravitational law, $g = GM/r^2$.

As stated previously, the simpler form of Newton's law is really all that is necessary to make everything work. This is so because once g's value is determined for any point from a body's center of mass, its only a matter then of multiplying g's value times any other mass in question that is being attracted by the attracting mass. For example, when we determine g at the Earth's surface as 9.8 N per kilogram and we wish to know its force of attraction on an average size human at its surface whose mass is approximately 70 kilograms, the total force is thus, 9.8 Nkg^{-1} times 70 kg equals 686 N. And since Newton's third law states that "for every force there is an *equal* but *opposite* force," then we know—without doing the calculation over just how much the 70 kg human body's force of attraction attracts the Earth to *it*. A 70 kg body therefore attracts the Earth's total mass of 6.0×10^{24}kg to it equally with a total force of 686 Newtons, and that is that. What one body's force is upon the other, the other body's force is equivalent, but *opposing* to the first. So, as one can clearly see, a housefly, an ant, an apple, your own body— they will attract the Earth with the same *precise* force that the Earth attracts them. That is the essence of Newton's third law, previously explained in more detail.

I had spoke previously of two one-kilogram masses positioned one-meter apart from center to center whose force between them is 6.672×10^{-11} Newton's—G's basic quantity. In reality, what this means as relates to a single one-kilogram mass (we will call it mass A) is that no matter how much smaller than a one meter radius mass "A" is in size; at precisely one meter from its center of mass in any direction from its surface, any other attracted body (we will call it mass B), will experience an *acceleration* towards "A" of 6.672×10^{-11}ms^{-2}. And if for some reason mass "B" was in orbit around "A", we would say that "B" experiences a gravitational force of attraction of 6.672×10^{-11}N if its mass were exactly one-kilogram. In other words, A's gravitational field strength, one meter away from its center of mass is 6.672×10^{-11}Nkg^{-1}. Therefore if B's mass, which is located one meter from "A" were *two* kilograms, it would experience a total force of 1.334×10^{-10} Newton due to A's pull on it. However, no matter what B's mass were to be, it would accelerate towards "A" at 6.672×10^{-11}ms^{-2}.

What if we now consider the acceleration rate of a freely falling particle due to gravity g at A's *surface,* in which "A" has a half-meter radius and a mass of one-kilogram? Using Newton's classical equation for gravity, $g = GM/r^2$, we will therefore have for g at A's surface

$$g = GM/r^2 \qquad (1-13)$$
$$g = (6.672 \times 10^{-11} \text{m}^3 \text{kg}^{-1} \text{sec}^{-2})(1.0 \text{kg})/(5.0 \times 10^{-1} \text{m})^2$$
$$g = 6.672 \times 10^{-11} \text{m}^3 \text{sec}^{-2}/2.5 \times 10^{-1} \text{m}^2$$
$$g = 2.668 \times 10^{-10} \text{ms}^{-2}.$$

Take note that g in equation (1-13) is $2.668 \times 10^{-10} \text{ms}^{-2}$, or equivalently, $2.688 \times 10^{-10} \text{Nkg}^{-1}$. Notice, however, that 2.688×10^{-10} is a value *very close* to the value of the New Gravity's a_u of $2.795 \times 10^{-10} \text{ms}^{-2}$. This a_u, I remind one, is the Universal Constant of Acceleration. Its value is a *direct linear constant of proportionality* between any body's *force* of gravity F_g at its surface, and the product of the body's column-*mass* m_c.

In addition to being a strict *ratio* between a body's gravitational *force* at its surface and the *mass* of its column, a_u's value will, with some detail, be shown in the following chapter to allegedly be the *same value* that Hubble's constant H_o is considered to be—at least as compares to H_o's current value by one of our leading and most honored astronomers, Allan Sandage. H_o's value represents the rate at which the universe is believed to be expanding. It will also be shown where a_u's value, and the theory itself behind it, have a direct connection with several other areas of physics where there are still unanswered questions.

Physical Parameters of the Fundamental Basic Sphere

In equation (1-13) above, it was shown that for the *basic fundamental* sphere having a *one-kilogram* mass with a *one-meter* or *less* diameter (in which two equal spheres of like dimensions were classically used to equate the quantity of force between them to be *one unit* of Newton's G) that g's value at the *surface* of one sphere having a half-meter radius was $2.668 \times 10^{-10}\text{ms}^{-2}$. As stated however, the mutual force between the two equal masses of one-kilogram each, positioned one-meter apart from their centers of mass, is precisely 6.672×10^{-11} Newton, *one* unit of G's value.

These facts are being restated in order to bring out an important distinction between Newton's constant G and the New Gravity's constant a_u.

That distinction has to do with this question: What, in fact, are the absolute precise physical parameters of nature's *basic* and *fundamental* sphere? In other words, when we utilize our own established set of physical units (the SI's MKS units) in conjunction with the basic equations we've constructed for gravity and gravitation, we should thereafter expect to find that there must indeed be some *one* specific spherical *mass* of some *one* specific spherical *radius* which ties in directly with the numerical value of $4\pi/3$.

This is being brought up as such because there does in fact appear to be just such a naturally basic and fundamental sphere as is herewith being suggested. The rationale for such a suggestion arose out of what was firmly believed to be a straightforward connection between the magnitude of the force of gravity at a body's surface and its column-mass.

It was found that for *each* kilogram of a body's column-mass (in essence, its density times its radius) the acceleration, due to gravity, of any freely falling body near the attracting body's surface was equal to $2.8 \times 10^{-10}\text{ms}^{-2}$. So that, for example, the Earth's 9.8ms^{-2} acceleration rate of freely falling bodies due to gravity near its surface, is strictly proportional to its column-mass of

35-billion kilograms multiplied by such proportionality constant, $2.8 \times 10^{-10} ms^{-2}$.

It was precisely and unequivocally this specific *column-mass* (the heretofore unrealized gravitational mass) which determined the precise amount of *gravity* at the surface of any body, whether we are speaking of protons and electrons or planets and stars.

The end result was a gravitational formulation which actually reflected Newton's second law, $F = ma$. The differences of course are F_g for the force of gravity; m_c for the *mass* of a body's one-square-meter column; and a_u to denote a universal *acceleration*.

What more could one ask for—a clean straightforward formulation of a *linear* nature having *less* hypotheses than the classical formulation of $g = GM/r^2$.

What are the magnitudes of these physical parameters which are born out of $4\pi/3$'s numerical value, 4.189? It struck me that since a_u's value, $2.8 \times 10^{-10} ms^{-2}$, was a strict ratio between a body's force at its surface and its column-mass that there had to be a connection between the fundamental *basic* sphere's size and density to $4\pi/3$'s quantity of 4.189; wherein F_g's value at the surface of this basic sphere would be precisely the same value as the ratio between it and the column-mass, m_c.

The rationale for thinking along these lines was also based upon the fact that since a_u's value turned out to be equal to G's value times $4\pi/3$; and because a_u was a *direct* constant of proportionality between a force and a mass in the gravitational context; and since the theory postulates that gravity is strictly due to acceleration; it then appeared natural that the equation for determining the *volume* of a sphere $(4\pi/3)r^3$, had to have a direct connection with a_u and F_g or F_o in which one unit of each would have to be equivalent, one to the other.

In order to uncover these so called *magical* parameters of radius, density and gravity, I first tried what some may believe is just plain numerology. It really was *not* numerology however since I firmly believed there was indeed a connection to be found between 2.8×10^{-10} and 4.189, other than a_u's value being the product of G and 4.189, which of course is the numerical value of $4\pi/3$.

I eventually determined that since the *basic fundamental sphere* (bfs) was of necessity to have a mass of *one-kilogram*, the *radius* was the only other key parameter that would be necessary to determine. However, I found the correct density first in an *unorthodox* manner. I took the square root of 4.189 and came out with 2.0467. This is *not* normally the formulation for density. The correct density formula is $\rho = M/V$. However, when I divided the 2.0467 kg/m^3 so-called density into the one-kilogram mass to determine the volume of the basic one-kilogram sphere, I arrived at 0.4886 m^3. This, it turned out, was precisely the correct volume and the correct radius (they each have precisely the same value). However, it still was not the orthodox manner in which to derive either density, radius, or volume.

In order to get more in line, I went after the *radius* in what seemed to be a more orthodox method. Since the formula for the volume of a sphere is $(4\pi/3)r^3$, I determined that the radius of the basic fundamental sphere (which was to cooperate with G, a_u, F_g, and F_o) should be calculable simply by deriving the *square root* of 1m^2 divided by $4\pi/3$. The quotient of 1m^2/4.189 is 0.2387m^2. The square root of 0.2387m^2 is 0.4886 meter, which in fact turned out to be the correct cooperating radius with the basic sphere's F_g and a_u, otherwise, 2.795×10^{-10}ms^{-2}. Thus, after determining the basic one-kilogram sphere's radius of 0.4886 m, its volume, very surprisingly, had the same numerical value as the radius (4.189 × 0.4886m × 0.4886m × 0.4886m = 0.4886m^3). This in itself was quite interesting to find that both the *radius* and the *volume* for the basic fundamental sphere—whose numerical values are born out of $4\pi/3$—are precisely the same values, 0.4886m and 0.4886m³, respectively!

Another duplicate value, it turned out, is the 1.0 kg *total* mass of the basic sphere, and the 1.0 kg gravitational mass of the basic sphere's *column*. Since the density ρ is in fact 2.0467 kg/m³, and the radius r is 0.4886m, the column mass m_c is the product of these and 1m²; so that we have for column-mass $(m_c = \rho r \cdot 1m^2) = 2.0467$ kg/m³·0.4886 m·1m² = 1.0kg).

I thereafter determined g's value for this basic fundamental sphere using Newton's classical equation, $g = GM/r^2$ with the basic 1.0 kg mass and the more natural 0.4886 meter radius. The result turned out precisely as I had guessed it should, thus

$$g = GM/r^2 \qquad\qquad (1-14)$$
$$g = (6.672 \times 10^{-11} Nm^2kg^{-2})(1.0kg)/(4.886 \times 10^{-1}m)^2$$
$$g = 6.672 \times 10^{-11} Nm^2kg^{-1}/2.387 \times 10^{-1}m^2$$
$$g = 2.795 \times 10^{-10} Nkg^{-1}.$$

Look back at equation (1–13). The basic value for g using Newton's half-meter diameter sphere is 2.6684×10^{-10}—not the critically needed ratio of a_u. Notice now that equation (1–14a) gives the same results as Newton's classical equation in (1–14); however, (1–14a) is strictly a *linear* equation with less hypotheses than (1–14), such that

$$F_g = m_c a_u \qquad\qquad (1-14a)$$
$$F_g = (1.0kg)(2.795 \times 10^{-10} ms^{-2})$$
$$F_g = 2.795 \times 10^{-10} N \text{ (no label of } kg^{-1}, \text{ only force, N)}.$$

As has been stated previously, all three (F_o, a_u, and F_g) have the *same* numerical value of 2.795×10^{-10}. F_o is the Outside Force pushing or otherwise acting upon the body thru space with a total force of 2.8×10^{-10} Newton times *each* kilogram derived from its column-mass; a_u is the Universal Constant of Acceleration of $2.8 \times 10^{-10} ms^{-2}$ which all bodies (proton or planet) undergo due to the *total* Outside Force acting on them; and F_g (g in Newton's equation with units of Nkg^{-1} or ms^{-2}) is the force of gravity at the surface, $2.8 \times 10^{-10} N$, or $2.8 \times 10^{-10} ms^{-2}$ times each kilogram of column-mass (m_c) which is imposed upon *each* kilogram of mass resting on a body's surface, or upon *any freely* falling object due to gravity near the body's surface.

To give an example of the above, the total Outside Force F_o acting upon, or otherwise pushing the Earth thru space, is $2.8 \times 10^{-10} Nkg^{-1}$ times $3.5 \times 10^{10} kg$ of column-mass, for a total Outside Force of 9.8 Newtons. This in turn causes the Earth to accelerate at a rate of a 9.8 N Force divided by a 3.5×10^{10} gravitational column-mass for an acceleration of $2.8 \times 10^{-10} ms^{-2}$. This acceleration in turn creates the force of gravity which pulls all other objects at its surface towards its center of mass by an equal but opposite force of 9.8 Newtons for each kilogram of any resting object's total mass; or which causes any and all freely falling objects near the Earth's surface to accelerate towards it at a rate of $9.8 \ ms^{-2}$.

Summarizing some of the more significant advantages the New Gravity has over Classical Gravitation we have:

1. The value arrived at for the Universal Constant of Acceleration a_u, appears to be an improved value over G's classical value, $6.672 \times 10^{-11} Nm^2kg^{-2}$, in that a_u's value, $2.795 \times 10^{-10} ms^{-2}$, is derived as a *direct* constant of proportionality between a *force* F_g and a *mass* m_c, thereby connecting Newton's 2nd law directly to gravity.

2. A body's Column-Mass m_c (which in *essence* is the product of its *density* and *radius*), appears to be the Gravitational Mass, meaning therefore, that

Inertial mass and Gravitational mass may not be proportional to one another after all.

3. The constant (a_u) is a Universal Constant of Acceleration of 2.8 Angstroms per second per second which is imposed upon every particle, body, etc., in the universe due to some Outside Force F_o which is acting or pushing on each massive object at the rate of 2.8×10^{-10} Newton for each calculable kilogram of a body's column-mass. This force then acts with just enough push on every particle, every body in the universe to cause each to accelerate at the constant rate of 2.8×0^{-10}ms^{-2}.

4. It is the product of the acceleration of a body thru space at that constant rate of 2.8×10^{-10}ms^{-2}, and the body's column-mass which creates the force of gravity at a body's surface; so that in effect F_g (g in Newton's Formulation) is always strictly and or linearly proportional to the calculable mass of the body's column.

5. Not only is a_u's value the product of G and $4\pi/3$ (whose 2.795×10^{-10} quantity is linear as in contrast to G's nonlinearity), a_u's value is also the magnitude of the force of gravity at the surface of the basic *fundamental* sphere whose size and density derive *directly* from the numerical value of $4\pi/3$'s (4.189).

6. The New Gravity teaches that a body's real volume is always hypothetically *Inflated* out to the point in space of the center of mass of the attracted body in question; and that a body's total inertial mass is always thought of as being *uniformly* distributed whether it is being hypothetically inflated or not.

7. Some interesting points which concern the *basic fundamental sphere* is that besides having both, a radius of 0.4886 meter, and a volume of the same

equivalent value, $0.4886m^3$, it also has both a total *inertial* mass of one kilogram, and a *gravitational* column-mass of one kilogram. Also, the ratio between its force of gravity and its column-mass is 2.795×10^{-10}—just as its force of gravity, F_g is also $2.795 \times 10^{-10}ms^{-2}$ and its total Outside Force is likewise $2.8 \times 10^{-10}N$. What is also very interesting is that the overall spherical area of this basic sphere with its 0.4886 meter radius is precisely three (3) square meters. All of this strongly indicates that the basic fundamental sphere is *no* accident.

8. The Column-Mass, in *essence,* is really determined as the product of a sphere's two key parameters— its *density* and its *radius.* As such, are we not really speaking of something of significance which agrees with the basic axiom of Newton's gravitational law, which is: *"The force between any two particles is an attraction along the line of joining them."* Is this not actually what we are speaking of—a *line,* the *radius* representing the *distance* from the *attracting* body's center of mass to the *attracted* body's center of mass?

Those eight items are only some of the changes and variations which differs from Newton's classical gravitation. Other items of significant interest were spoken of prior to this point in the book; and additional items, especially those which are indirectly related to gravity will be brought out, especially as they relate to Unifying Gravitational Interactions with Electromagnetic Interactions.

The values for physical constants and parameters have generally been shown to three decimal places. Some, such as those which stem *directly* from π can naturally be shown to any degree of precision one desires to know. Since π's value (3.1416) is the *most precise* value we know of in physics, the values for the physical parameters of the basic *fundamental sphere,* which do in fact stem from the value of $4\pi/3$, will be shown up to 12 decimal places in the table which follows—Table 1–1.

The Universal Constant of Acceleration (a_u) will be shown to six places in table 1–1 by multiplying G's best known value of 6.67259×10^{-11} times $4\pi/3$ for a value of 2.795008×10^{-10}. Following this list of constants and physical parameters there will be discussion at length on an item of much *curiosity,* in which there appears to be some grounds for determining the precise values of G and a_u (not from experiment, but from theory) to a degree of precision matching that of π's precision.

Can a_u's Precise Value be Some Fraction of π?

Speculating on a value for a_u and G to a degree of precision *equal* to that of pi (π) is what the last three items in the physical constants and parameters column are about.

It is—crazy as it sounds—an attempt to tie in a_u's precise value, and thereby, Newton's Gravitational Constant G with π. Naturally, if this can be done, a_u's and G's value can be determined to an infinite number of decimal places, i.e., with all the precision of pi.

What would be the rationale to even begin to think that there may in fact be a direct relation between π's value and the precise values for a_u and G?

For starters, it has been made clear throughout that:

(A) a_u's value is shown to be a *direct linear constant* of *proportionality* between the force of gravity at a body's surface, F_g, and the body's Column-Mass m_c; such column-mass being the *gravitational* mass of a column whose *volume* has a one-square

$G = gr^2/M$	Universal Gravitational Constant	G	6.672 59x10^{-11}m^{3}kg^{-1}sec^{-2}
	(Theoretical)	G	6.666 666 ...x10^{-11}m^{3}kg^{-1}sec^{-2}
$a_u = F_g/m_c$	Universal Constant of Acceleration	a_u	2.795 008x10^{-10}ms^{-2}
		or	2.795 008x10^{-10}ms^{-2}kg^{-1}m$_c^2$
	(Theoretical)	a_u	2.792 526 803 16x10^{-10}ms^{-2}
$F_o = F_g/m_c$	Universal Outside Force	F_o	2.795 008x10^{-10}N
	(Theoretical)	F_o	2.792 526x10^{-10}N
$g = GM/r^2$	Acceleration Due to Gravity Near Earth's Surface	g (Classical)	9.806 65 ms^{-2}
$F_g = m_c a_u$	Acceleration Due to Gravity Near Earth's Surface	F_g (New Gravity)	9.806 65 kgms^{-2}
$m_c = \rho r \cdot 1m^2$	Gravitational (Column) Mass of Basic Sphere	m_c	1.0 kg
$M = (4\pi/3)r^3\rho$	Inertial (total) Mass of Basic Sphere	M	1.0 kg
$\sqrt{1m^2/(4\pi/3)}$	Radius of Basic Sphere	$r\sqrt{1m^2/(4\pi/3)}$	0.488 602 511 906 m
$V = (4\pi/3)r^3$	Volume of Basic Sphere	V	0.488 602 511 906 m^3
$\sqrt{4\pi/3}$	Density of Basic Sphere	ρ (unorthodox)	2.046 653 415 89 kg/m^3
$\rho = M/V$	Density of Basic Sphere	ρ	2.046 653 415 89 kg/m^3
$4\pi \cdot r^2$	Area of Basic Sphere	A	3.000 000 000 000 m^2
$F_g = m_c a_u$	Force of Gravity at Surface of Basic Sphere	F_g	2.795 008x10^{-10}N
	(Theoretical)	F_g	2.792 526 803 16x10^{-10}N
$F_g = m_c a_u$	Acceleration Due to Gravity at Surface of Basic Sphere	F_g	2.795 008x10^{-10}kgms^{-2}
π	(pi)	π	3.141 592 653 59
$4\pi/3$	$12\pi/9$	$4\pi/3$	4.188 790 204 77
$8\pi/9$	Theoretical Universal Constant of Acceleration	a_u*	2.792 526 803 16 Å/s^2
$G = a_u/(4\pi/3)$	Theoretical Gravitational Constant (Classical)	G*	0.666 666 666 66...Å
$8\pi/9 \div 12\pi/9$	Theoretical Gravitational Constant (Classical)	G*	0.666 666 666 66...Å

*Elaboration on these theoretical values immediately follows.

Table 1–1. Some of the Constants and Physical Parameters as Used in This Book THE NEW GRAVITY.

meter cross section and a length of the body's radius, or distance to an attracted body's center of mass.

(B) G's value ties in *directly* with $(4\pi/3)$, in that their *product* is a_u's value, 2.795×10^{-10} ms^{-2}kg^{-1}m^2, or ms^{-2}.

(C) Most significantly, it has been shown that the gravitational field strength, otherwise the force of gravity F_g at the surface of the *basic fundamental* sphere, is in fact precisely *one unit*, 2.795×10^{-10} Newton of force, which in turn acts upon each kilogram of mass resting on its surface. This is the product of a_u's value, otherwise the Universal Constant of Acceleration of 2.795×10^{-10}ms^{-2} and the one (1.0) kilogram of gravitational mass calculable from the *basic* sphere's *column*. It is also the *same* value of one (1.0) basic unit of the alleged Outside Force F_o, whose value, 2.795×10^{-10}N (applied to each kilogram of a body's column mass) is the precise numerical value of a_u's 2.795×10^{-10}ms^{-2}.

(D) It has clearly been shown that the basic fundamental sphere's physical parameters: i.e., its mass, volume, density, and radius are *all* tied in *directly* with, and *clearly stem* from $(4\pi/3)$'s value of 4.188 790 204 77.

The exceedingly interesting matter of the *numerical* value of π and $4\pi/3$ being directly tied together in so many ways with G's 6.672×10^{-11} and a_u's strict ratio of 2.795×10^{-10}-to-one between the gravitational *force* F_g and the gravitational column-*mass* m_c, is that this ratio is also extremely close to the 2.792 526 803 16 value of $8\pi/9$. Here then is the *kicker:* $8\pi/9 \div 4\pi/3$ or $12\pi/9 = 0.666\ 666\ 666\ 666...$! In essence, G's value may very likely be 0.666 666... Angstrom units. As an aside, $8\pi/9$ has other applications in physics.

Once again then, the question is: "What would be the *rationale* for one to suspect that a_u's value, and therefore G's value, can numerically stem from π and $4\pi/3$, respectively.

Let's begin by looking at the history of G and its value. The best numerical value for G to date is $6.672\ 59 \times 10^{-11}$. Its value is the least accurately known of *all* the physical constants of nature since it is really only known to four decimal places.

Down through the years G's numerical value has been found by laboratory experiments to be anywhere between 6.465×10^{-11} in 1881 by Jolly, and 6.754×10^{-11} by Cavendish in 1798. The present value of 6.672×10^{-11} was established by Heyl and Chrzanowski in 1942 and has not been found to differ to date.

Since G is still the *least* accurately known of all the physical constants, this leaves room for a very interesting theory. That is that its absolute value appears to be the quotient of $8\pi/9 \div 4\pi/3$ which, once again, is 0.666 666 666 666...? In other words, since a_u's value is the product of G's value of 6.672×10^{-11} and $4\pi/3$ for a product of 2.795 Angstrom units; then, numerically speaking, why not a_u's value being a slightly adjusted G value of 0.666 666 666 666... times $12\pi/9$ (otherwise $4\pi/3$) for a product exactly equal to $8\pi/9$—a precise value of 2.792 526 803 16 *Angstrom* units?

It should not be too surprising that G's absolute numerical value might indeed stem from the precise value of π, in that it can possibly be the quotient between $8\pi/9 \div 12\pi/9$. After all, are not many areas of physics laden with formulas utilizing π's value or multiples of π in so many different ways?

What happens if we *reduce* the size of the basic fundamental sphere by ten powers of (10) to a *radius* of 0.4886

Angstroms (0.4886×10^{-10}m). It's volume would then be 0.4886Å^3 (0.4886×10^{-30}m^3); its mass would be (1.0×10^{-30}kg); its column-mass (1.0×10^{-10}kg); its over-all spherical area (3.0×10^{-20}m^2); and its density as originally stated, 2.046×10^0kg/m^3. The value for F_g, the force of gravity at its surface, would then be 2.795×10^{-20}N; however a_u would remain as 2.795×10^{-10}ms^{-2} times each kilogram of column-mass.

Utilizing these extremely small dimensions of one Angstrom unit or less instead of meters, one can now consider π's value of 3.1416 as more accommodatingly having a direct relation to G and a_u's value, and therefore show their relation even more clearly. This relation is the adjusted value for G of 0.666 666 666 666... Angstrom unit; and for a_u is the adjusted value of 2.792 526 803 16 Angstroms. Here again, a_u's 2.792 526 803 16—otherwise $8\pi/9$'s value—is exactly equal to the product of G's 0.666 666 666 666... and $4\pi/3$.

Are We in Fact Accelerating Thru Space?

One of the most significant facts about the New Gravity is that it reveals to us that it is we—the Earth, Moon, Sun, and the rest of our Milky Way galaxy—that are actually propagating thru space at the *phenomenal* speed we attribute to light, notably, 186,282 miles (299,792 kilometers) per second!

What does all of this reveal? To begin with, it reveals that if in fact we here on Earth, and the Earth itself, are indeed moving thru space with a *present* velocity of 186,282 miles per second, it would necessarily be due to the fact that all the particles which the Earth, ourselves, and the other astronomical bodies, etc., are comprised of, have been *accelerating* at the constant rate of

$2.8 \times 10^{-10} \text{ms}^{-2}$ since time zero $t = 0$, where it is generally believed the universe had its origins.

To determine the amount of time t the Earth's particles, and of course the rest of our Milky Way galaxy, have been accelerating since $t = 0$, we need only divide our present velocity thru space—which in essence is the present velocity of light c—by the Universal Constant of Acceleration a_u, such that we would calculate for our own age

$$t_{MW} = c/a_u \qquad (1-15)$$
$$t_{MW} = 2.998 \times 10^8 \text{ms}^{-1}/2.795 \times 10^{-10} \text{ms}^{-2}$$
$$t_{MW} = 1.073 \times 10^{18} \text{sec}$$

and since a year consists of 3.157×10^7 seconds, our age (A) in *years* would simply be 1.073×10^{18} seconds divided by 3.157×10^7 sec/yr; thus our age at this particular epoch of space-time would be

$$A_{MW} = t_{MW}/\text{sec yr}^{-1} \qquad (1-16)$$
$$A_{MW} = 1.073 \times 10^{18} \text{sec}/3.157 \times 10^7 \text{sec yr}^{-1}$$
$$A_{MW} = 3.4 \times 10^{10} \text{yrs.}$$

Notice that (A_{MW}) in equation $(1-16)$ is *not* light years but, in fact, *years*. Light years reflect *distance*. If we desire to know the *distance* d in which the particles comprising the Milky Way galaxy have actually traveled since $t = 0$, we would then have to cut our present velocity of light c in *half* for an *average* velocity over time of $1.499 \times 10^8 \text{ms}^{-1}$. This would therefore account for the fact that we have been constantly and eternally *accelerating* for a time of 34 billion *years* at a rate of $2.795 \times 10^{-10} \text{ms}^{-2}$. Thus for the distance traversed since time-zero, we would have traveled at an average velocity of $1.499 \times 10^8 \text{ms}^{-1}$ for a time t of $1.073 \times 10^{18} \text{sec}$. We would therefore have for *our* galaxy's particles total distance traveled since $t = 0$ to the present

$d_{MW} = (1/2c)t_{MW}/m$ ly^{-1} (meters per light year)   (1–17)
$d_{MW} = (1.499 \times 10^8 ms^{-1})(1.073 \times 10^{18} sec)/$
        $9.461 \times 10^{15} m$ ly^{-1}
$d_{MW} = 1.608 \times 10^{26} m/9.461 \times 10^{15} m$ ly^{-1}
$d_{MW} = 1.7 \times 10^{10}$ ly.

This of course is 17 billion *light years;* the age has *always* been expressed in light years because astrophysicists have never related the age of the universe by any other scale than light years. That is so because it was not considered that light had any other velocity throughout space and time than what we presently know it to be—299,792 km per sec. Our age therefore, was *likewise* stated as 17 billion *years.* However, if we are in fact accelerating for *all* time, our age would naturally have to be twice our distance in light years. This will be borne out in much detail in Chapter 8.

What is being advocated in this book is that it is we ourselves—our planet Earth and the rest of the Milky Way galaxy—along with all the other galaxies of the universe that are physically moving through space at the velocity we attribute to light; and that Gravity itself is the direct result of this constant acceleration through space.

That acceleration of 2.8×10^{-10} meter per second per second is such an infinitesimal amount (only one-ninety-millionth of an inch per second per second) that it may be incomprehensible for one to envision that such enormous and massive bodies as stars like the Sun, and planets like the Earth could effectively and eternally be accelerating this very minuscule distance (equal to the diameter of an air molecule) for each second of time as they fly off through space at their present velocity of

186,282 miles (299,792 km) per second. From all the facts available however, this seems to be just the picture.

Universal gravity may well then be attributable to all bodies inherently accelerating at a constant rate of 2.8 Angstroms per second per second, and Electromagnetic phenomena would then be directly attributable to a body's present (basic) velocity through space. Thus when we compare the Electromagnetic interactions to the very minuscule Gravitational interaction, *both of which appear to derive from the same basic mechanism,* we do indeed come up with the ratio R, as so often has been shown to exist between the two—a ratio of 10^{36}-to-one. For when one divides the square of this Universal Constant of Acceleration a_u^2, into the square of our present velocity of light, c^2, *each for an elapsed time of one-second,* we find that the Electromagnetic interaction is indeed a *trillion-trillion-trillion,* times more powerful than the Gravitational interaction, both being a *direct* consequence of the Universal Constant of Acceleration, thus we have for this critical ratio, R

$$R = c^2/(a_u \cdot 1\sec)^2 \qquad (1-18)$$
$$R = 8.988 \times 10^{16}/7.812 \times 10^{-20}$$
$$R = 1.151 \times 10^{36}.$$

Equation (1–18) appears then to provide the relationship much sought after between Electromagnetic interactions and Gravitational interactions, both being directly related to the Universal Constant of Acceleration a_u, and our own present velocity through space. Consequently, this ratio (10^{36} to one) should grow larger with time since a_u remains constant while c constantly increases with time by 2.8 Angstroms per second per second. At present then, the approximate value of c at our space-time epoch of the universe would be determined by the product of the length of time t a body has been accelerating since $t = 0$ (1.073×10^{18} seconds in our own case) and body's rate of acceleration a_u, such that for the Milky Way MW, we would have a present velocity for c of

$$c_{mW} = a_u t_{MW} \qquad (1-19)$$
$$c_{MW} = (2.795 \times 10^{-10} ms^{-2})(1.073 \times 10^{18} sec)$$
$$c_{MW} = 2.999 \times 10^8 ms^{-1}$$

and for our own velocity V_{MW} here at the Milky Way position in space-time, we would do exactly as shown in equation (1–19), so that

$$V_{MW} = a_u t_{MW} \qquad (1-20)$$
$$V_{MW} = (2.795 \times 10^{-10} ms^{-2})(1.073 \times 10^{18} sec)$$
$$V_{MW} = 2.999 \times 10^8 ms^{-1}.$$

Those two equations (redundant as they are) naturally determine both, the real velocity *of* Earth through space and Earth's local velocity of light which emanates from Earth's real velocity at this particular epoch of space-time in relation to the beginning of the universe's expansion at $t = 0$.

There are still many areas yet to be covered in the New Gravity. Its implications will get into areas later in the book such as the *quantum dilemmas* of the infamous *wave-particle dualism* and the electron's *momentum-position* riddle. It will get into the explanation for Einstein's famous equation on energy, $E = mc^2$, and its connection with the Universal Constant of Acceleration. It will attempt to explain why light propagates with one constant velocity; and will show why a_u's value and H_o's value (Hubble's Constant) are allegedly one and the same quantity; plus a host of other important areas relating to the physical universe on a *large* scale, and the physical universe on the scale of the *atom* and its constituent parts, down into the realm of the *quantum*.

The Physical Units of Newton's Gravitation and the New Gravity

One final item of discussion in this chapter upon which I wish to elaborate are the *physical units* of Newton's *classical* gravitation and the *New Gravity*.

As stated earlier, G's physical units (Nm^2kg^{-2} and $m^3kg^{-1}sec^{-2}$) are said to be peculiar to G, but were of necessity *chosen* as such in order that G would be consistent with Newton's Universal Law of Gravitation. This law is specifically formulated as $F = GM_1M_2/r^2$.

When I initially began to tackle this area of physics, I can recall on several occasions inquiring with many physics professors, as to what does *Newton-square meter* (Nm^2) and *square-kilogram* (kg^{-2}) mean in Newton's *"Force"* equation for gravitation? I would also follow up with "and what does cubic meter per kilogram (m^2kg^{-1}) mean in Newton's simpler form of the equation for determining both, the acceleration due to gravity *near* a body's surface, and the *gravitational* field strength at its surface in which both are determined with $g = GM/r^2$?

In each case of asking the above questions I always received the same response: A *smile* and, *"Oh! Those units were only used to make G work with Newton's law of gravitation."*

Actually, there are very few statements of any detailed information throughout the physics texts and other literature concerning the history and mechanics of G. I personally dug deep to find more on G (everything about it) but failed to come up with much. Some would say, "What's there to come up with—its this and its that and that is it!" However, that just is not the case.

What "G" *originally* was in fact—and still is in essence—is a constant *force* (F) of 6.672×10^{-11} *Newton* (N) which is multiplied times *each* kilogram of the product of two masses in Newton's equation, $F = GM_1M_2/r^2$; or each kilogram of the single mass in his simpler form of equation, to express a body's force of gravity F_g, or otherwise, its gravitational field strength, $g = GM/r^2$. In this simpler and *more practical* form, g equals G's value of 6.672×10^{-11} meter per second per second (ms^{-2}), or 6.672×10^{-11} Newton per kilogram (Nkg^{-1}) for each kilogram of the single body's mass divided by the square of its radius (r), or otherwise, the square of the distance from the single body's center of mass to the center of mass of any attracted body in space.

As can be seen from its description, G actually has the unit of *Force,* and therefore, the unit denoting it would be (N) for Newton of Force. But notice that Newton's principal equation expressing the Universal Gravitation, $F = GM_1M_2/r^2$, have following G a first mass and a second mass, each expressed as (kg) times (kg), followed by dividing the product of G in Newton's (N) and the two masses (kg^{-2}) by the square of (r) which is denoted as meters (m) times meters (m).

What do we now have in all of the above? If we lay it out just as I have described it, we have in effect: $F = N \cdot kg \cdot kg/m \cdot m$. Since G originally was denoted as a force F in the unit of Newton (N), then if we leave it as such we end up with $F = Nkg^{-2}/m^2$. That does not net us (N) for Newton of Force as F for force demands! In order then to have the equation $F = GM_1M_2/r^2$ net us a value for F in Newtons only, what had to be done, *by necessity,* was to add to G's naturally flowing unit of Newton of Force "N", a square kilogram (kg^{-2}) for the two masses in the equation, and a square meter (m^2) for the two lengths of squared radius on the right side of the equation.

So then, by adding the (2) kgs and the (2) meters to G's unit (N) we finally have a set of units that will cancel out properly in order for F in Newton's universal gravitational equation to be denoted as it should, simply in Newton's (N).

In other words, G's natural and single unit of "N" will not net F with "N" units as it demands. However, G's *chosen* additional units of (m^2) and (kg^{-2}) now cancel out the two kilograms and two meters on the left and right side of $F = GM_1M_2/r^2$, and therefore satisfy the requirement that F's value result in the single unit of Newton (N) only.

With Newton's simpler form of the equation describing "gravity," and *not* gravitation, (there *is* a difference, technically speaking) this simpler equation: $g = GM/r^2$ has in it only one-mass M. This equation is the off shoot from $F = GM_1M_2/r^2$ as can readily be observed. However, $g = GM/r^2$ is generally applied when one needs to determine the acceleration due to gravity near a body's surface. With the Earth, for example, we would find g's value simply by multiplying G's 6.672×10^{-11} value times the Earth's 6.0×10^{24}kg mass and divide their product by the product of the Earth's radius of 6.371×10^6m times 6.371×10^6m. The net result would be $9.8ms^{-2}$ because in this particular case G's units would once again not have attached to it, its natural unit of $(kgms^{-2})$ which is the equivalent of (N). In this case G was given an additional unit of square-meter (m^2) in order that g would result in (ms^{-2}). G therefore had to be fitted up with the units: $m^3kg^{-1}sec^{-2}$. In this manner g ends up having the necessary unit (ms^{-2}) describing the *acceleration* required of it.

There is yet one additional item to be explained. When it is required that we need to know how much a body at the surface on the Earth weighs, we still utilize Newton's simpler form of his gravitational law: $g = GM/r^2$. Here again, if we apply to G in this case the original set of units, Nm^2kg^{-2}, we end up with 9.8 Newton's per kilogram (Nkg^{-1}). However, whenever it comes out as Nkg^{-1} we can still call it ms^{-2}; and if we end up with ms^{-2} we can call it Nkg^{-1} since one is the equivalent of the other.

For the sake of additional clarity to what has been stated, I will list here this equivalency of Newton's (N), kilograms (kg), and meters per second per second (ms^{-2}):

$$1N = 1kg \ m/sec^2$$
$$1kg = 1N/(m/sec^2)$$
$$1Nkg^{-1} = 1m/sec^2.$$

Now, when the *chosen additional* units for G are applied as the units for the New Gravity's a_u, we will find that they describe how a_u functions, thus if we want to determine g for the Earth, we have

$$g = \rho r a_u \qquad (1-21)$$
$$g = (5.516 \times 10^3 kg/m^3)(6.371 \times 10^6 m)$$
$$\quad (2.8 \times 10^{-10} ms^{-2} \ kg^{-1} m_c^2)$$
$$g = 9.8 ms^{-2}.$$

So that in effect, we find that g's acceleration near the Earth's surface is equal to $2.8 \times 10^{-10} (ms^{-2})$ for each kilogram (kg) derived from its square-meter column (m_c^2). Therefore, equation (1–21) is a naturally defining equation in which one can have a clear mental construct of what, in fact, the units represent. These units, by the way, $ms^{-2} kg^{-1} m_c^2$ are the equivalent of Newton's original arrangement of $m^3 kg^{-1} sec^{-2}$ in which $m^3 kg^{-1}$ has no logical connection to anything, as such.

To demonstrate the use of G's units once again using them with a_u in determining the mutual force of gravitation between the Earth and Moon, we have

$$F = \rho ra_u M_M \qquad (1-22)$$
$$F = (2.512 \times 10^{-2} k/m^3)(3.844 \times 10^8 m)$$
$$\quad (2.8 \times 10^{-10} Nm_c^2 kg^{-2})(7.345 \times 10^{22} kg)$$
$$F = 1.98 \times 10^{20} N.$$

Here once again, with the variation of using Nm^2kg^{-2} instead of $m^3kg^{-1}sec^{-2}$ for a_u's units, we define a_u as: 2.8×10^{-10} *Newton* (N) times each *kilogram* (kg) of the hypothetically inflated *square-meter column* (m_c^2) times each *kilogram* (kg) of the Moon's total mass.

Although equations (1–21) and (1–22) function satisfactorily with a_u using G's units, I did not choose to utilize that formulation in the "New Gravity" because there was no clear cut way to derive from them the sole unit of ms^{-2} to accommodate what a_u truly represented—an *acceleration*.

I instead decided to go with a_u as constructed from the formulation of $a_u = F_g/m_c$ in which F_g denotes the *force* of gravity at a body's surface in Newtons (N) only, and *not* Nkg^{-1} as does "g" with Newton's classical equation: $g = GM/r^2$. The symbol m_c in the New Gravity's key equation, $a_u = F_g/m_c$, represents the body's gravitational column-mass (m for mass and the subscript c for column. The physical unit for m_c is the kilogram for the *column mass* it represents. This gives to a_u the resultant unit of Nkg^{-1}, otherwise, its equivalent of ms^{-2}, denoting the acceleration that a_u truly is.

Back in February 1989 I mailed an eight page paper to 11,500 United States and Canadian physicists. The paper was about the theory which this book subscribes to. The physical units for a_u came out just as G's units do, $m^3kg^{-1}sec^{-2}$. I received, as a result, many letters from physicists who received my paper, pointing out that a_u was not in fact an *acceleration* as I had indicated it was. Their argument was simply that the units, "$m^3kg^{-1}sec^{-2}$"

is not an acceleration. I later placed—in the May '89 issue of *Physics Today*—a correction on this matter with a revised equation structured as: $a_u = F/m_r$, which is almost exactly as the New Gravity's equation: $a_u = F_g/m_c$. The m_r denoted the "Mass of one Square Meter Radius" and was further defined as a *new unit* (actually, it was a new physical parameter with a new unit) which reflects a body's Radius in *Cubic* Meters rather than its customary radius in *meters;* and whose mass is in fact a spherical body's Gravitational Mass.

I bring this matter up because it has indeed been somewhat of a problem since day one to end up with everything in the line of *physical units* just the way I'd have liked. However, just as Newton's G had to have some help for its resultant units to fit his law of gravitation, the law of gravity—as proposed in this book—also required some slight adjustment with its resultant units for F_g. As has been shown many times throughout, F_g's resultant unit is simply "N" for *Newton* of force. The equivalent of F_g is "g" in Newton's equation, which has the resultant unit of Nkg^{-1}. The "per kilogram" is really a *label* explaining how "Newtons of force" are applied; i.e., g's Newton force is multiplied by *each* kilogram of mass it acts on. To get around this omission of "per kg" with F_g's resultant unit of (N), F_g is defined only as: Newtons of *force* at the attracting point in question. It will be understood, however, that this magnitude of force will apply to each kilogram of *any* attracted body. Likewise, when F_g is alternatively meant to indicate an "acceleration due to gravity," the Newton (N) has the classically arranged equivalent unit of kg ms^{-2}. As is seen, this is not technically ms^{-2} as units of an acceleration normally result in. To repeat then, we end up with F_g having units of kg ms^{-2} where g has units of ms^{-2}; and units of Newtons where g has units of Nkg^{-1}. And where G has units of m^{3}kg^{-1}sec^{-2}, a_u has units of ms^{-2}. G's units, recall, were chosen as such (and therefore deemed peculiar) so that G would be consistent with Newton's Law of Gravitation. However, those same units of G are *natural,* and *not* chosen units for The New Gravity's a_u, which is $2.8 \times 10^{-10} ms^{-2} kg^{-1} m_c^2$. Remember too, a_u's value of $2.8 \times 10^{-10} ms^{-2}$ is simply the product of G's $6.67 \times 10^{-11} Nkg^{-1}$ and $4\pi/3$! In essence, Nkg^{-1} is the *only* unit G's value really reflects; for as stated earlier, $(m^2 kg^{-1})$ were *chosen* additional units in order that G's value would cooperate with Newton's Law of Universal Gravitation.

2

The Philocity of Light

Since the main thrust of the theory this book subscribes to is about both, the Philosophy of light, *and* the Velocity of light, I felt compelled to invent its connotation—PHI-LOCITY.

How many mysteries of the universe might be resolved if it were realized that the particles comprising our galaxy (electrons, protons, neutrons, etc.), have been accelerating through space in one general direction at a constant rate of 2.8 Angstroms per second per second for a distance since t = 0 of 17 billion *light years* over the past *34 billion years,* and are presently up to a speed of 186,282 miles per second? That's right, the speed of light!

Indeed, if this is true, then both the propagation of light itself and its precise speed would be direct consequences of the velocity of solid bodies through space; and as these solid bodies—such as our own Earth—continue to accelerate to higher speeds, ever so slowly, so too would the speed of light, *mimicking precisely the velocity of the Earth through space.*

And what of the mystery of Gravity? Would it not finally show itself for what it really is? For if in fact the elementary particles which make up the Earth and Galaxies, etc., are constantly accelerating through space at 2.8 Angstroms per second per second, then that *"acceleration itself"* must surely be responsible for the *"gravitational attraction"* that holds us to the Earth like a magnet.

How can we find justification for all of this? One key to it is wrapped up in the knowledge which we presently possess of an Expanding Universe, for there is no better telltale sign to indicate that the Earth and all other material bodies in the universe are constantly accelerating through space, than the direct evidence—first discovered by Edwin Hubble in 1929—that the universe is in a state of constant, and apparently, *uniform expansion.*

The Connection Between the Expanding Universe and Universal Acceleration a_u

Hubble's value H_o for the rate of expansion of the universe was originally estimated to be about 550 kilometers per second per megaparsec. This is about 170 kilometers (100 miles) per second per million light years. However, its present revised value is now only about 10% of his original estimate and is believed to be somewhere nearer to 18 kilometers (11 miles) per second per million light years, or 57 kilometers per second per megaparsec, of which the megaparsec—3.2 *million light years*—is the unit of distance on the cosmological scale preferred by astronomers.

Now if we were to work this rate of expansion of 18 km per second per million light years backwards towards *smaller* and *smaller* increments of space—instead of towards higher and higher increments as is usually the case—then, when we finally get as close in to only *one* light second (186,282 miles), the expansion rate would actually work out to be approximately 5.6×10^{-10} meter (5.6 Angstroms) per second per light second. However, since it is here advocated that the velocity of light is NOT after all constant, but instead *constantly increases with time*, then naturally, the light year distance scale must also increase with time so that the Hubble constant of 18 km per second per million light-years should, in reality, only be *one-half* its estimated value, or about 9 kilometers per second per million light years, which therefore reduces down to 2.8×10^{-10} meter (2.8 Angstroms) per second per light second.

In any case, we will concentrate on the recessional velocity of Hubble's constant for one light second—186,282 miles in distance—ignoring for the moment that recessional velocities are generally believed to apply only between clusters of galaxies, and here on in refer to Hubble's constant as having a value of 2.8 Angstroms per second per light second, as it will be shown that a_u and H_o are one and the same value.

Now since I have contended that gravity is strictly the result of a body's constant acceleration of 2.8 Angstroms per second per second through space, it must then be shown how this acceleration through space is actually related to the Hubble constant H_o.

My own adjusted value for the Hubble constant of 9 kilometers per second per million light years (which as stated works out to about 2.8×10^{-10} meter per second per *light* second) would be the actual required value in order for the force of gravity F_g near the surface of the Earth to equal 9.8 Newtons; or the Moon to equal 1.6 Newtons, etc. etc. Thus, this constant, a_u, accounts for both the *Force of Gravity* (F_g) on the surface of all material bodies throughout the universe, and the *Outside Force* (F_o)

pushing these bodies thru space—whether we are speaking of electrons and protons or planets and stars.

Other Mysteries Resolved with a_u

What other mysteries of the universe might be resolved if indeed our galaxy, the Milky Way, is advancing through space at a speed heretofore strictly attributable to light and other forms of electromagnetic phenomena?

If light does indeed derive its energy of motion from the *motion of the source* it is emitted from, this would then answer for certain the question as to why *nothing can ever exceed the speed of light*. The answer, naturally, would be evident. It is because light simply *mimics* the velocity of anything that moves. Move faster—and light moves faster! Light and other forms of electromagnetic interactions are energy carrying images; and images consist of photons, and therefore, they have no mass. Consequently, they actually shoot away with *real* velocities and or *apparent* velocities strictly proportional to that of the body they are emitted from. In our case here on Earth, that velocity would of course be 299,792 km sec^{-1}; or in British units, 186,000 miles per second.

Additional mysteries which can be cleared up if indeed the Earth is rushing off through space at the speed we attribute to light, would be in connection with Einstein's famous equation on *mass* and *energy*, $E = mc^2$. And of course, the big question: *Why does light propagate with one specific speed?* Answers to these two questions then become immediately apparent. The velocity of light c in the equation $E = mc^2$ is squared simply because this velocity is actually the *Real Velocity v of the Mass m* for which we are determining the rest energy E.

As for why light always appears to propagate at one specific speed, would that not be apparent if indeed it is we who are rushing off through space at the speed we attribute to light? In our own case here on the Earth, we and the Earth rush through space at a speed of 186,282 miles per second, although all along we are constantly picking up speed—accelerating at the minuscule rate of 2.8 Angstroms for each second of time. However, for all practical purposes, it would still be correct to state that light propagates with one specific speed—always at the speed with which we here on Earth and the Earth itself rush off through space—186,282 miles per second. But that is only for this particular moment of time. During the next second of time, our speed increases by about one ninety-millionth of an inch (2.8 Angstroms), and for each additional second of time our speed increases another 2.8 Angstroms, and so on and so on, ad infinitum. However, small as this acceleration rate is, wherein the Earth's velocity thru space only amounts to an increase of about *one centimeter per year,* it has accelerated each and every particle which comprises our galaxy to a velocity of 186,282 miles per second over the past 34 billion years (10^{18} seconds) at an eternally constant rate of acceleration. For this reason, the Milky Way's distance from $t = 0$ is presently 17 billion light years.

Technically speaking, there cannot be any such thing in nature then as a *perfectly constant velocity!* For all practical purposes however, we can safely state that light always propagates with one specific speed, i.e., with one constant velocity.

3

The Not So
Constant Constancy
of Light's Velocity

What do we actually mean when we speak of the constant velocity of light? Webster's definition of "constant" means *not to change, i.e., remaining the same.* The prime theme of this book is that the velocity of light is indeed NOT constant, and that the only thing constant about it is its *constantly increasing velocity* by 2.8 Angstroms per second per second.

There are generally three modes in which the constant velocity of light is viewed throughout the literature:

First, there is "constant" in the sense that light always propagates with one specific speed. In this sense of "constant," light never has to accelerate from zero velocity (i.e., from its presumed position of rest) to get up to its one speed of propagation of 186,282 miles per second. This evidently is what James Clerk Maxwell meant in teaching that the velocity of light was constant. He presumably had no knowledge of whether or not the velocity of light was an absolutely unchanging constant, for even with our present day technology it is very doubtful that we would be able to detect such a minuscule change in the velocity of light by only 2.8 Angstroms per second. This is only one ninety-millionth of an inch per second per second.

The main thing to be remembered when we speak of this Constancy of lights velocity is that this is indeed, for all practical purposes, an accurate statement as such, and it should not be allowed to become entangled with other versions of the word "constant" in connection with light. Maxwell's meaning was and still is very correct. He stated in effect, that *light always moves from place to place with one speed!* Now if light does indeed increase its speed by only *one part in one-thousand-quadrillion per second,* he certainly had no idea of this, and he certainly wasn't implying anything of this nature when he said what he did about the way in which light propagates. He was simply referring to the fact that light's velocity (of the order of 3.0×10^8 meters per second) was the only velocity with which it moves about from place to place. That of course implied it did not have to "accelerate" in order to achieve that speed. It was, in other words, a *built in speed.*

The Failed Michelson-Morley Mode

A *second* mode in which light is referred to as moving from place to place with a constant velocity, is that which refers to the famous experiments of the 1880's by Michelson and Morley in which they set out to detect a variation of light's velocity due to the Earth's 18 mile per second orbit around the Sun.

It is well known that Michelson and Morley failed to detect the expected variation in the velocity of light in

accordance with "The Principle of the Addition of Velocities." Their experiments showed instead that no matter
which direction they pointed their light beam in respect
to the Earth's forward motion around the Sun, the velocity of light did not appear to show any variation whatsoever. As a result of their very meticulous experiments, it
was then presumed that the velocity of light was *also constant in this other sense,* which, in effect, is quite different
from that which Maxwell had first postulated.

So even in this sense Michelson and Morley's experiments showed (in yet an other mode) that light always
had the same speed no matter "which direction" the
beam was aimed from a *moving* body. In other words, and
according to Einstein, even if the moving body—the
Earth in this example—were traveling around the Sun at
99% the speed of light, the results Michelson and Morley
had gotten of the Earth's 18 mile per second orbit around
the Sun would be the same result they would get if the
Earth could indeed orbit the Sun at 99% the speed of
light (184,419 miles per second instead of 18 miles per
second). This type of reasoning provides a good example
of what Einstein most likely meant when he denounced
the use of *common sense* in contemplating the mechanism
of the universe.

What Michelson and Morley's experiments *supposedly*
proved then, was not only did light move about from
place to place at one specific speed as taught by Maxwell,
but that the additional velocity of 18 miles per second
supposedly imparted to the light beam from the Earth's
orbiting of the Sun, did not make any difference whatsoever in the beam's rate of propagation. It neither gained
in relative speed when aimed in the Earth's forward direction around the Sun, nor lost in relative speed when
aimed at right angles to the Earth's direction around the
Sun. The Principle of the Addition of Velocities *did not
appear* to apply in the case of "light propagation."

So here we have a second version of what the Constant Velocity of Light actually means. I intend to show however that there is a good *common sense* explanation for the results which Michelson and Morley had gotten; and that the Principle of the Addition of Velocities *does* in fact hold true with respect to light.

Einstein's Absolute Limiting Velocity of Light

Now for the third version of the Constant Velocity of Light—*Einstein's version*. This is the version which I suspect may have brought physics to somewhat of a standstill.

After Michelson and Morley's failure to detect a variation of the velocity of light, Einstein, professing not to have known of their work, supposedly played with the idea for some years as to why light's velocity did not adhere to the rules of the Principle of the Addition of Velocities. Finally, in the early part of 1900 Einstein made some far reaching assumptions about this failure of light's velocity to vary. What he said was that the reason light's velocity did not vary in those cases where it was expected to vary, that is, in accordance with the Principle of the Addition of Velocities, was because there was No Ether in the first place to carry the light. The *ether* was at that time believed to pervade all of space and was thought to be the medium for carrying the light waves—just as *air* is the medium for carrying *sound waves*. Secondly, and most importantly, Einstein said that *The Velocity of Light was an Absolute and Unchanging Constant Throughout All Space and Time*. He said in fact, it was *space* and *time* which were suspect, which themselves were *not* constant; and that our concept of space and time simply was not accurate.

What Einstein was saying in effect, was that the speed of light *never* changes anytime, nor anywhere throughout the vast universe—not even by one-angstrom unit per year—not by any amount throughout *all* space and time!

There was no argument that light always propagates with only one speed; nor any argument that it did *not have to accelerate* to get to its sole propagating speed of 186,282 miles per second. This being so, and in addition to the fact that the Michelson-Morley experiments failed to detect any change in the speed of light when common sense dictated that they should, Einstein finally reasoned that common sense is not always the answer. He stated that past experiences tend to prejudice our thinking. This then is when he decided that our concepts of *space* and *time* must not be totally accurate, henceforth, came his Special Theory of Relativity in 1905 in which he put a "Limiting Velocity" on the speed of light, and therefore, as I have stated previously, he may have *"unwittingly"* steered the physics community slightly off course.

The important factors then, which Einstein felt were absolutely necessary if he were going to be able to change the way we think about space and time, was that not only does the velocity of light always propagate at one specific speed without having to accelerate to get to that speed— but that the velocity of light was also an *absolute and unchanging constant throughout all space and time.* So that no matter how fast, nor in which direction an observer or a light source moved with respect to one another, the Principle of the Addition of Velocities was not, he said, not applicable to light as it was with other phenomena of nature. This thinking, which I purposefully repeat once again, arose strictly as a consequence of his assumption, which he later raised to a *postulate,* that the speed of light was an Absolutely Unchanging Constant of Nature. So that even if one could move alongside a light beam at 99% the speed of light they would measure that light beam to be traveling 186,282 miles per second away from them, and not the 1,863 miles per second away from them as the principle of the addition of velocities would naturally and normally require.

This then concludes the different interpretations when referring to the Constancy of light's velocity.

4

The Three Truths

$E = mc^2$, The Constancy of Light's Velocity, and The Expanding Universe

It is true indeed that the speed of light is constant, that is, constant in the sense that it does in fact propagate from place to place with one specific velocity. However, I will attempt to show that the speed of light is *not,* I repeat, is *not* an absolutely unchanging constant of nature.

In addition, I will attempt to show why Michelson and Morley always found the velocity of light to be the same regardless of the direction they pointed their light beams.

Finally, I will attempt to show why the *rest energy* E in Einstein's famous equation, $E = mc^2$, is always equal to mass m times the velocity of light squared c^2.

The big questions then are: "Why does light not have to accelerate to get up to speed? And secondly, "Why does it travel at the particular speed that it does" (186,282 miles per second) and not say, 400,000 miles per second, nor 100,000 miles per second?

The answers to these questions can be found bound up in *Three Basic Truths*—three facts of life which scientists have so diligently uncovered.

$$E = mc^2$$

The first of these Three Truths will be that which concerns Einstein's famous equation on mass and energy, $E = mc^2$.

What does $E = mc^2$ actually mean in plain and simple terms? It simply means that the "rest energy" of each and every piece of matter in the universe is equal to the product of a body's mass and the velocity of light squared. Now I ask, how much less complicated can nature be? For what this actually means is that the rest energy which is bound up in everything and anything here on Earth (including the Earth itself) is equal to 90 quadrillion (9.0×10^{16}) joules of energy for each and every kilogram of matter.

Take for example a lump of coal on Earth weighing 2.2 pounds. This is equal to a mass of approximately one kilogram. Or take a rock having this same mass of one kilogram, or a piece of metal of equal mass. Whichever material we use, it makes no difference. What $E = mc^2$ teaches is that the rest energy bound up in any of these one-kilogram chunks of matter is equal to the same amount of energy required to electrically heat one million homes for one year, based on an average home using 25,000 kw hours per year for heating. This is derived at using Einstein's famous $E = mc^2$ simply by multiplying the coal, the rock, or the metal's mass of one kilogram by the velocity of light squared, which is about 300,000,000 meters per second squared. The squared product then is 9.0×10^{16} joules (a joule being a unit of energy). And since one kilowatt-hour is equal to 3,600,000 joules, we find there is an equivalent of 25-billion kilowatt hours of energy bound up in any one kilogram piece of matter in our piece of the universe.

Hypothetically speaking, this enormous amount of rest energy should also be equivalent to what would happen if you could hit the Moon from the Earth with a 2.2 lb. (one kilogram) rock in one and a quarter seconds after being released. Of course, the velocity of the rock would have to be an *unaccelerated* constant velocity of 186,282 miles per second from the very moment it left your hand (ridiculous as that sounds) to the very moment it hit the Moon. The resultant energy released upon impact would be tremendous. Theoretically, it too should be equal to the energy of 25-billion kilowatt hours of electricity, just as $E = mc^2$ predicts.

This then is how nature can provide us with a *sensible* and *logical* explanation for such a phenomenal amount of stored energy bound up in such small amounts of matter. There can only be one explanation for this phenomenon, and that is that the Earth—at its present position in space and time—is itself rushing off through space at a present speed of 186,282 miles per second, so that in effect, $E = mc^2$ is actually $E = mv^2$. In reality, this would mean that the rest energy equals the mass times the Earth's own real velocity squared. This being the case, we ourselves (including the coal, the rock, the Earth, etc.) are the real factors possessing the basic velocity which must be squared! Henceforth, both the propagation and the precise speed of light's propagation would follow as a direct consequence of the Earth's (or any other body's) basic velocity through space. Therefore, $E = mc^2$ lays itself bare to "reason" after all.

Einstein himself did not claim to "basically" know why his famous energy equation worked, but $E = mc^2$ did work, and it worked well. Here then, in $E = mc^2$, lurks a piece of solid evidence in support of an uncomplicated model for the universe, a piece of evidence, *which by itself*, should suffice to reveal to us the true secret of Light's Flight throughout the universe!

The Constancy of Light's Velocity

The *second* of the Three Truths I refer to states that, *"the velocity of light is always constant."* Now certainly, if it is indeed a fact of life that $E = mc^2$ is proof in itself that the Earth is rushing off through space at a present speed of 186,282 miles per second, it should then become crystal clear why the velocity of light (for all practical purposes) propagates with one specific velocity. I say, "for all practical purposes," because as I believe, light's velocity will be found to be steadily increasing by approximately 2.8 Angstroms per second per second as a result of that *same increase* in velocity of the body it is emitted from, that is, from the Earth and everything attached to it.

The most amazing thing however about the *not so constant constancy of light's velocity,* is that when a light source is activated, such as, say a flashlight, its beam never starts out from zero-velocity as would logically be expected of anything which is presumed to be at rest rather, its beam starts out *unaccelerated* with a starting or initial velocity of 186,282 miles per second!

Now common sense and logic, contrary to those who hold that such niceties do not always prevail, tells us that *nothing can just up and go on a journey without having to start out!* Yet we know it to be an indisputable fact that a light beam does just that—it always begins its journey with an initial speed of 186,282 miles per second from a position which is "presumed" to be at rest. This is absolutely impossible in the realm of everything we know and see around us, yet that is precisely what light *appears* to do! Whatever happened to that part of its velocity range from zero through 186,282 miles per second? How does it manage to skip the natural stage of acceleration and just happen to begin its journey with a velocity of 186,282 miles per second? The only explanation which can realistically support the nature of this *constancy of light's velocity* is the same as that given as to why $E = mc^2$, and that is that the materials which the Earth and everything upon it are made (including the flashlight in this example) have in fact been on a journey for billions of years (34 billion to be exact), constantly accelerating at 2.8 Angstroms per second per second, and are now presently traveling through space at the enormous speed of 186,282 miles per second. In other words, it is only logical to suspect that before switches are pulled, or buttons are pushed to activate a light source, that the light source itself (again the flashlight in this case) *must already had been in motion,* moving along with the Earth at 186,282 miles per second, and that the *unaccelerated* propagation of its beam of light at its one and only speed of 186,282 mps is simply a direct consequence of its present basic velocity through space in the first place.

In other words, the flashlight's cold (unenergized) image was *already* propagating at 186,282 mps: However, by energizing the flashlight we simply *enhance its image* by heating the filament of its lamp, and reflecting a multiplicity of bright white filament images, such that it then becomes readily visible when impinging (as it already had been, when unenergized) upon opaque materials. It is analogous to a high velocity stream of water rushing past you. If you pour some red dye, for example, into the speeding stream, the moment the dye makes contact with the stream, it immediately takes on its velocity (without having to accelerate), i.e., if the stream's velocity is a hundred feet per second, the dye's *initial unaccelerated* velocity is also a hundred feet per second. This then is the case with the light beam from the flashlight. In effect, we changed the color (temperature), and therefore the *brightness* of the flashlight's *filament* in mid stream, i.e., the flashlight's cool dull streaming image was already moving at c before altering the filament's image.

Here also, as with $E = mc^2$, Einstein did not claim to know why light propagated with one specific velocity. No one had every explained why light propagated with one velocity, it remained a mystery. However, it was this Constancy of the Speed of Light which played the major role in Einstein's Special Theory of Relativity. And though it was not actually he, but James Clerk Maxwell who first made it known that the speed of light always propagated with one specific velocity, Einstein made use of Maxwell's, work entitled "The Nature of Light," and in 1905 followed up with his Special Theory of Relativity. In this, he indicated that not only does the speed of light always propagate with one specific velocity, but that its speed was also an Absolute and Unchanging Constant of Nature Throughout All Space and Time.

Here then, Einstein made a whole new assumption when he said the velocity of light throughout the universe of space and time was an absolute and unchanging constant. All the usual predictions of his Special Theory of Relativity then followed, both, as a consequence of his four-dimensional Space Time Continuum, and his assumption that the velocity of light was the *maximum and limiting velocity*—the same unchanging velocity for all observers at all times—past, present, and future, throughout the universe.

It was here then where it appears to be that the "monkey wrench" may have gotten thrown into the mechanism, because it was this *absolute and unchanging value of light's velocity* which thereafter became "A Way of Life." It became a widely accepted postulate of the Nature of Light throughout the scientific world, and therefore had most probably brought whatever real progress was being made to somewhat of a standstill.

If, on the other hand, it had been realized that the velocity of light was NOT an absolutely unchanging constant, but *does in fact increase with time*, it might then had become more likely to find answers to questions such as: Will the universe go on expanding forever? How can the four forces of nature be unified—specifically, *gravity; the electromagnetic interactions; the strong nuclear force; and the weak nuclear force?*

Of course, if the Earth *is* in fact moving through space at a present velocity of 186,282 miles per second, and *if* indeed it is constantly accelerating at a rate of 2.8 Angstroms per second per second, then it should be relatively easy to calculate the magnitude of g at its surface.

Einstein did in fact teach that the effects of *acceleration* and *gravity* are, for all practical purposes, *indistinguishable* from one another. In that case, if the Earth is constantly accelerating at a rate of 2.8 Angstroms per second per second, then the magnitude of F_g, the force in Newtons at its surface, should strictly be due to the effects created from the minuscule rate of acceleration by more or less the second law's equation ($F_g = m_c a_u$), where m_c represents a body's column-mass, i.e., its mean density ρ times its mean radius r times one-square-meter $1m^2$—such *column mass* defined herein to be the body's Gravitational Mass, and of course, a_u in this new formulation representing the Universal Constant of Acceleration.

Hubble's Expanding Universe

In addition to Earth's minuscule rate of acceleration being responsible for creating its gravitational field, we should then know the Hubble constant more precisely.

This constant should be found to be in the range of 8 to 9 kilometers per second per million light years as presently appears to be the case. Technically however, it will not appear constant much beyond 3 or 4 billion light years simply because "constant acceleration" can only be equated with a constant which is associated with *time*, and not *distance*, as the Hubble constant connotes. And so, the Hubble constant is not constant *even* up to 3 billion light years, but only appears constant.

As for the issue which concerns the Deceleration Parameter q_o, there should be no question that there is *no* deceleration—*no slowing of the expansion* if the galaxies are constantly accelerating at the rate of 2.8 Angstroms per second per second. This should instead be regarded as strict *uniform expansion* since the rate of expansion is constant for *all time* throughout the whole of the universe. It is strictly a *linear expansion* then when viewed as a *velocity-time* relationship, and *non-linear* when viewed as a *velocity-distance* relationship. For in any accelerating system of bodies where a strict velocity-time relationship exists, it will be found that the further ahead in time and space we measure for recessional velocity between any two bodies at *equidistant* points, the lesser will the recessional velocities be between them as they increase their distance per unit time. As a result, it is plain to see that recessional velocities *per unit distance* in an accelerating system, will *decrease* as time goes on, thereby imparting a somewhat *false* sense of a *deceleration effect* when looked at from a velocity-*distance* point of view.

In contrast to this, if we were able to view this expansion process of galaxies in the opposite direction (that is inwardly, in the direction of t = 0) we should find as we look further and further behind us—so to speak—that there exists just the opposite effect: the galaxies (per unit distance) recede from each other *faster* and *faster* as we look further and further *back* towards the center—towards the younger galaxies—where the expansion process supposedly began. This viewpoint would likewise impart a "false sense" of a *speeding up* of the expansion, when in fact it would really be an indicator of *true* uniform expansion.

What we then have is that when we detect deceleration ahead of us, that is, away from $t = 0$ in the expanding system per unit distance, and acceleration behind us in the direction of $t = 0$ in the expanding system per unit distance, these are only an assurance that *strict* uniform expansion is indeed taking place. In other words, there is *no slowing whatsoever* of the acceleration process going on in the universe. It is indeed expanding, but the fact is that there can be no real Hubble constant in relation to velocity and *distance* over the long run: a true constancy of expansion can only apply to a "velocity-time" relationship, as is depicted in Table 8–1, the Universe in Numbers.

In this, the third of the Three Truths, the universe *is* indeed in a state of *uniform expansion*. Although it was Slipher who originally showed that the galaxies are receding from us, it was through the very meticulous work of the American astronomer, Edwin Hubble, that the secret of uniform expansion was revealed.

Hubble's discovery centered on the fact that the distant clusters of galaxies were receding (moving away) from us and each other at velocities which appeared to him to be directly proportional to their distance. Hubble arrived at this conclusion for the simple reason that when he found, for example, a cluster of galaxies to be a million light years away, their recessional velocity was determined by him to be approximately 100 miles per second; and when he found galactic clusters 2 million light years away (twice the distance as the first) their recessional velocity was found to be 200 miles per second—twice that of the first. If he detected galaxies at a distance of 20 million light years, they were measured to be receding at 2,000 miles per second. (Note: Velocity-Distances scale is now estimated to be only about 10% of Hubble's original estimate.)

Hubble supposedly found that same pattern no matter how far out in space he looked. He consistently found that the recessional velocity was apparently in direct proportion to distance, and therefore took this to mean that the universe was expanding in a uniform manner.

If we compare this uniform expansion, for example, with recessional velocities and expansion rates between solid bodies falling towards Earth, it will be found that except for the fact that all galaxies *diverge* from each other, whilst a multiplicity of falling bodies towards Earth's surface *converge* and fall towards Earth's center of mass, there is not the slightest difference in the structure of the equations describing their motion, except for their particular constants of acceleration of 9.8 meters per second per second for Earth bound objects, and 2.8 Angstroms per second per second for space bound galaxies.

The method Hubble used to make his far reaching determinations comprised the use of the well known Doppler effect. With the Doppler effect the wavelength of light is shifted toward the blue end of the spectrum (high frequency, shorter wavelength) when a source of light is moving towards an observer. When a source of light is moving away from an observer, the wavelength of the light is shifted towards the red end of the spectrum (lower frequency, longer wavelength) in which the waves appear to be spread out. Also, with the Doppler effect, the greater the speed the greater the shift in wavelength. Therefore, by determining the extent of displacement in the spectral lines, it then becomes relatively easy to determine the recessional velocity between the observer and the source. Once this recessional velocity is known, it then becomes a matter of simple mathematics to estimate the distance of galaxies because the velocity-distance relationship appeared to Hubble to be in direct proportion to each.

What does this expansion process mean? First, I might comment that Hubble's discovery that the universe is expanding in an orderly fashion was the most important cosmological discovery ever made!

Originally, Hubble estimated the recessional velocity between galactic clusters to be about 100 miles per second per million light years (our present light year is about 5.9 trillion miles—the distance traversed by light at the rate of 186,282 miles per second during a time span of one year). However, after many years of investigation of the velocity-distance scale by others, the constant has now been revised all the way down to approximately 11 miles per second per million light years, although it is still disputed. This latest refinement of the Hubble constant is the result of many years of work by astronomers such as the distinguished Allan Sandage. It is a number which I have found to be in excellent agreement with my own equation for determining the Universal Constant of Acceleration.

Now although the accuracy of the Hubble constant is extremely critical—both for the purpose of determining a_u and for the purpose of determining, more precisely, the age of our own galaxy—the real significance of Hubble's discovery is the fact that the universe of galaxies are indeed expanding uniformly, that is, they are spreading apart from each other in an orderly fashion which is commonly referred to as the Hubble law. Hubble's law simply states that the further away a galaxy is, the faster it is receding from us.

In order to understand what is actually taking place in a uniformly expanding universe, I can think of no better way to describe it than to relate it to something we are

familiar with here on Earth—and that is the influence
which gravity has upon bodies in free fall as relates to
rate of fall and distance traveled. In fact, it will be seen
that the *same* physical laws which apply to freely falling
bodies under the influence of *gravity* here on Earth, are
precisely the *same* physical laws which apply to the *expansion* of the universe.

The only differences to be found between these two
accelerating systems will be that one is a *convergent* system,
while the other is a *divergent* system. And other than their
rates of acceleration being radically different, they both
obey the same physical laws. Let us take for example a
quantity of golf balls held at rest about 5,00 feet above
the Earth's surface, then release them. What happens
naturally, is that they would all descend from their positions of rest and converge towards the center of mass of
the Earth as they accelerate at a constant rate of speed
of 32 feet (9.8 meters) per second per second. The only
difference between *this* system and the *expanding* universe
of galaxies is that the galaxies all move outward, *diverging*
instead from each other through space, and accelerate at
a constant rate of 2.8 Angstroms per second per second
(about one ninety millionth of an inch per second per
second) as in contrast to 32 feet per second per second
for the converging golf balls. Thus when the particles
which comprise the galaxies go forward at a_u's constant
rate of acceleration of 2.8 Angstroms per second per second, they will attain a speed of 186,282 miles per second
at the end of 34 billion years, just as the particles of our
own galaxy has, and will have traversed a distance from
$t = 0$—in one general direction through space—of 17 billion of our present light year of 5.9 trillion miles;
whereas, when the golf balls fall toward the Earth's surface at a constant rate of acceleration of 32 feet per second per second, they will be traveling at a speed of 320

feet per second at the end of 10 seconds, and will have traveled a total distance of 1,600 feet.

To elaborate further on these two systems, let us say that the whole Earth is hypothetically covered over by a dozen or so layers of golf balls situated about a mile above its surface. Each layer is spaced, say ten feet above the other, and the golf balls in each layer are spaced about 300 feet apart. Now then, all the balls in the first layer surrounding the Earth are released *simultaneously,* and in each ensuing second of time another layer of balls are released until in twelve seconds time all twelve layers of balls are released.

Let us also suppose, for the sake of argument, that all the balls continue to accelerate at a constant rate of 32 feet per second per second (ignoring the fact that the Earth is a solid body) and they fall 4,000 miles in 19 minutes to the very center of mass of the Earth where they will impact with a final velocity of 7 miles per second— converging into a somewhat hot solid mass.

As can be seen, the only difference between the two systems are that the golf balls accelerate inwardly to eventually converge upon each other at a rate of 32 feet per second per second, while the galaxies accelerate outwardly in a *divergent* fashion through an infinite space *away* from each other at only 28. Angstroms per second per second.

Of course, there is no comparison in their respective rates of acceleration since the golf ball's acceleration rate of 9.8 meters per second per second near Earth's surface is approximately 35 billion times greater than the acceleration rate of the Earth and Galaxies through space of only 2.8×10^{-10} meter per second per second.

In each of these situations however, both the golf balls and the galaxies *double their velocity at four times the distance traveled.*

As can be seen then, there is no difference whatsoever in the principal laws which govern the motions of accelerating bodies—whether they are galaxies accelerating through infinite space in a *divergent* fashion, or golf balls accelerating to Earth through a finite space in a *convergent* fashion. In any case, each system accelerates in strict accordance with the same basic laws of nature.

5

"Why" the Velocity of Light is Not Constant

In the last chapter I had spoken mostly in general terms about the three most tested, most accepted, and most important cosmological discoveries of our time. These are: *Einstein's energy equation E = mc²; the constancy of the velocity of light,* and the discovery of a *uniformly expanding universe*—uniform in the sense that it is a *linear* type velocity-distance expansion. I of course advocate a *non-linear* velocity-distance expansion since I have stated previously the strict proportionality of a velocity-*time* relationship.

It is my opinion that each of these three discoveries can stand alone to support what I have stated over and over, that is: "that it is we ourselves, the Earth and the rest of the galaxy, that are accelerating through space at a present velocity of 186,282 mps, and that both the propagation and the precise speed of light are direct consequences of our own basic motion through space."

Why wasn't this conclusion reached a long time ago? The answer to this is not a simple one, and may not even be a nice one. I say this only because if what has been concluded in this book is indeed correct about the basic mechanism of light, then it may have to bear upon Einstein's Theory of Relativity; for although Einstein contributed *many* good and positive works in addition to his energy equation, it was however he who *solidly* established the postulate that *light propagated with one absolute and unchanging velocity,* i.e., that its speed was one and the same speed in the *past,* is the same at *present,* and will forever remain the same in the *future*—not only here in our vicinity of space—but throughout the whole of the universe.

Now certainly, it is not debatable to say that light normally propagates at *one* specific speed. However, if we instead say that light propagates with only one *absolute* and *unchanging* speed throughout all space and time (as Einstein had postulated) it then becomes a debatable issue! Still however, there is nothing wrong if we say light propagates with "one constant velocity," that is to say, "constant for all practical purposes"; for if the only change in its approximate velocity of 186,282 miles per second amounts to an increase of, *not* kilometers per second, *not* meters per second, *not* even centimeters or millimeters per second, but only approximately Three-Angstroms, or one ninety-millionth of an inch per second, then we can correctly state for all practical purposes that light does propagate from place to place with a constant and or unvarying velocity.

So there we have it, the velocity of light is, *for all practical purposes,* constant, just as Einstein and Maxwell before him taught. However, Einstein apparently had not realized that one very important characteristic about light, and that is its very small, very minuscule and constant increase in its speed by approximately 2.8 Angstroms per second per second. Granted, that is only an increase in speed for each second of time of *one part in a thousand-quadrillion* of its present speed, but it is in fact an increase, insignificant as it may seem. It brought the Milky Way's particles 17-billion light years away from time-zero over a time span of 34-billion years at that rate.

Indeed, it is this small but constant increase which is the only factor in the expansion of the universe. The propagation of light itself would appear then to be a *secondary* effect arising from the basic motion of the astronomical bodies thru space.

As for Gravity, there should be no question of its basic mechanism once it is fully realized that the massive Earth, for instance, is moving through space with a constant acceleration of 2.8 Angstroms per second per second, just as every other body in the universe is doing.

As stated earlier, Einstein showed that the effects of acceleration and gravitation are *indistinguishable* from each other. Therefore, if indeed the Earth *is* accelerating, we should confidently be able to state that the force of gravity F_g on Earth is solely attributable to the fact that it is presently speeding through space with a constant acceleration of approximately 2.8×10^{-10} meters per second per second.

As for finding a direct relationship between the forces of Gravity and the Electromagnetic interactions, it would seem evident what that connection is! It would allegedly seem that the gravitational force should be due to the Earth's *acceleration* through space at a constant rate of 2.8×10^{-10} meters per second per second; and that the precise velocity of light should then be due to the Earth's present velocity through space of 186,282 miles per second after steadily accelerating for 10^{18} seconds, otherwise, 34 billion years thru 17 billion light years of space.

I am not aware of any existing technology which affords us the capability of detecting an increase in the velocity of light by as little as *three Angstroms per second* (about one centimeter per year). If such technology could afford us this capability, then of course we would be in an excellent position to test this theory which, in effect, would be a direct method to determine both, our basic velocity through space, and our basic acceleration rate through space by using light itself as the measuring rod.

This of course is what Einstein's Special Theory of Relativity runs directly contrary to, whereas he himself had decreed that "It is futile to try to determine the true velocity of any system by using light as a *measuring* rod."

So once again, I repeat: It is this absolute and unchanging value of the speed of light which Einstein decreed to be a fact of life which I feel may have slowed progress in physics.

I have wondered for many years now why, with so much solid evidence as the Three Truths confront us with, why we had not come to the conclusion a long time ago that: *The basic motion which determines the speed of light is the same basic motion of the planets and stars.* I suppose the front runner of any argument to that question might well be that which Einstein himself had postulated when he said the velocity of light is an absolute and unchanging constant of nature throughout all space and time.

It is of course true that light does indeed propagate with one speed! What this means however is that light does *not* have to go through the process of *accelerating* from zero velocity, i.e., from rest, to get up to its present speed of 186,282 miles per second. We can safely state that much about light since that much about the character of light is proven beyond any doubt. (Note: Please overlook, if you can, the repetition in making my point; it is simply *too important* not to be repetitive.)

To go on then, think about this if you will: Think about the way in which light propagates at the colossal speed of 186,282 miles per second and *never even has to start out to attain that speed!* Ponder upon this for awhile and then ask yourself: How can anything in nature just simply up and take off with an initial speed of 186,282 miles per second from a position of presumed rest *without first having to accelerate* to get to that speed? First of all, it must be realized that one factor is clear and indisputable—and that is that light does indeed always propagate in empty space with *one* constant velocity, 186,282 mps at the present time, however little it may increase from second to second as the theory herein postulates.

The point then is this: Since there is no dispute whatsoever of that one particular characteristic of light, i.e., it doesn't have to accelerate to get up to speed, there should then be no argument that the source which emits the light must itself had *already been in motion at that particular speed,* and for that reason, and that reason alone, light's built in instantaneous velocity is what it is.

I believe that when the idea first emerged of the "constancy of light's velocity" from Maxwell's equations, it was only intended to be construed that it did not have to accelerate up through the velocity range from zero to 186,282 mps to get to 186,282 mps, but instead always propagated at one instantaneous *built-in* speed for whatever the reasons. Einstein evidently concluded that this constancy of the velocity of light should also represent that its velocity never *throughout all space and time,* changed even infinitesimally; and so, to my knowledge, it was Einstein who put the real emphasis on the velocity of light being a limiting velocity—an absolutely unchanging velocity for all time and space throughout the universe.

Ironically, it was Einstein himself who may have put the one obstacle in his own path during the last 30 years of his life which he spent trying to Unify Gravity and Electromagnetism but failed.

6

"Why" $E = mc^2$

As I have contended all along, we certainly must already be in motion through space, racing along with a present speed of 186,282 mps in order for light and all other forms of electromagnetic radiation to leave their source with a *built-in* starting speed of 186,282 mps.

I do not feel we should need any additional evidence to convince us of our actual flight through space at the speed we attribute to light. However, I take up once more the important discovery which Einstein himself made concerning Mass and Energy. There is no dispute of the validity of Einstein's equation on mass and energy, $E = mc^2$. In the previous sections I have explained what this equation meant. You will recall that I used as an example a lump of coal, a rock, or a piece of steel, each having a mass of one kilogram, and illustrated how and why, according to Einstein's famous equation on mass and energy, they each possess their enormous stores of energy.

Einstein himself proclaimed that he did not know why such a phenomenal store of energy resided in all material bodies attached to the Earth, including the Earth itself.

Imagine what this so called rest energy would be for the Earth as a whole? To envision this, consider that the Earth is rushing through space as I propose it to be, at a speed of 186,282 mps. Let us say that it is heading for the Sun, and the Sun is instead a cold dark star having a somewhat similar density and or material make-up as the Earth. Let us also say the Sun is not moving, that it is just suspended there in space. Now imagine the Earth running head on into the Sun at a speed of 186,282 mps. The total release of energy at that moment would actually be equal to the exact amount of energy which $E = mc^2$ predicts the Earth presently possesses—about 5.4×10^{41} joules. The same applies to the Sun, and according to $E = mc^2$ its rest energy amounts to 1.8×10^{47} joules. However, this stored or rest energy would be exactly equal to that which would be released if the Sun instead were to run head on into a stationary solid body (similar to the Sun itself) at a speed of 186,282 mps.

Imagine if you will, two hot flaming stars such as the Sun, each rushing towards each other from the outer edges of two different expanding universes. Each of these two stars then collide with the other head on at a speed of say 300,000 kilometers per second, if not more. The energy and total light output from two hot stars such as the Sun colliding head on at 300,000 km per second each should, in theory, be equal to what is happening when we are looking at Quasars; for if what I have proposed in this book are facts to be reckoned with, then there may be good reason to believe that Quasars are simply *two colliding stars* near the outer edges of our own expanding universe and a second expanding universe, both of whose outer edges have eventually come to expand into each others turf.

Getting back however to the philosophy of $E = mc^2$, as I have pointed out over and over, $E = mc^2$ is one of the Three Truths I refer to. The equation itself is not generally, if at all, in dispute. Einstein had given to the world a great contribution in $E = mc^2$. I ask then of the reader, once again, to ponder upon this famous equation. Ponder its implications, consider how significant it has been to science, and how it stands today upon solid ground as to its validity.

Now since no one has yet explained precisely *why* $E = mc^2$ works, and since there is no dispute whatsoever that light does indeed propagate at one speed; and considering my own reason why light only propagates with one speed; how else then can a mass of only one kilogram (2.2 pounds of weight on the Earth) store such an enormous amount of energy, *energy sufficient to heat a million homes for one year,* if it were not for the plain and simple fact that the one-kilogram mass is itself speeding along with the Earth through space at the phenomenal speed of 186,282 mps?

This then seems like the only rational explanation why Einstein's energy equation calls for multiplying the mass by the velocity of light squared. It is—in reality—the velocity of the coal, rock, steel, etc., which is actually being squared, so that in essence, *the equation is really* $E = mv^2$.

So unless we decide that common sense and logic should be ignored, the velocity of light squared in Einstein's famous equation ($E = mc^2$) seems clearly to stand for the actual squared velocity of the object of mass in question; and so, the energy (E) would in effect actually be *Kinetic Energy!* In this case the object is a rock, a lump of coal, an automobile, or the Earth itself, all of which have a basic velocity of 186,282 mps by virtue of their present space-time position along with the rest of the Milky Way galaxy.

This then, very simply, is how I believe *nature* has managed to give to light waves, radio waves, and all forms of electromagnetic phenomena, a vehicle with which to move from place to place. As such, Einstein's Principle of the Addition of Velocities would apply to light after all. However, our measurements may never provide us with anything except one detectable velocity for all observers. See for example, Figs. 9–1 through 9–11 in Chapter 9, "On the Way Light Propagates."

7

"Why" the Universe Expands

So much said for $E = mc^2$ and the Constancy of the Velocity of Light. If these two truths should not suffice to convince one that they are traveling along with the Earth at the speed we attribute to light, then Hubble's discovery of a Uniformly Expanding Universe should provide a third piece of evidence.

The most significant factor about the Expanding Universe centers upon Hubble's Constant, H_o. This constant provides us with, not only the rate at which the Universe expands, but put another way, *the rate of universal acceleration through space of all matter throughout the universe.*

In other words, since the present estimated value of the Hubble constant is approximately 18 kilometers per second per million light years, then it is only reasonable to work this figure backwards, that is, closer in to our own galaxy, all the way in to a distance from Earth of only *one light second,* 299,792 km or 186,282 miles. When this is finally reduced from 18 kilometers per second per million light years all the way down to only one light second, the resultant figure is about 5.6×10^{-10} meter per second per light second, otherwise 5.6 Angstroms per second per light second. In this manner, we could presume that the Earth and all other matter in the universe is accelerating through space at a constant rate of 5.6 Angstroms per second per second.

As I indicated throughout the book however, the actual *near field* constant should eventually be found to be about 9 kilometers per second per million light years (instead of 18 kilometers) since the velocity of light is *"not"* after all constant. This of course will then fit in with a Universal Constant of Acceleration of approximately $2.8 \times 10^{-10}ms^{-2}$ which is the actual ratio between the force of gravity F_g at the surface of all material bodies throughout the universe, and the product of their mean density and mean radius times one-square meter.

Consequently, what we presently believe to see at one million light years away should actually prove to be two million light years away, and so on. Therefore, 2.8 Angstroms would then be appropriate, as this would reflect a *real* Hubble constant of 5.5 miles per second per million light years or 28.7 km per second per megaparsec instead of 57.4 km per second per megaparsec.

Hubble's universe is one of simplicity and beauty; however, we seem to have found a way to complicate the picture. Let me explain why. If one were to indulge in, say *fifty*, or even a *hundred* books related to cosmology, it would easily be concluded that the universe is basically very simple in its structure—and rightly so. Hubble's discovery revealed that not only was the universe expanding, but that all galactic clusters were moving away from each other at speeds which were in direct proportion to their distance from each.

Now no matter which of the literature one indulges in on this subject they will find that all or most give the same interpretation of the Expanding Universe. This interpretation generally states that *all* clusters of galaxies in the universe are moving away from each other. They will generally teach—and rightly so—that as time goes on, all galactic clusters move further and further apart, and that this expansion process has been going on for close to 20-billion years (my estimate is 34-billion years at our own space-time epoch). They will also indicate that the size of the universe back about 20 billion years ago when it first began to grow, was extremely small, and at that particular time in the past when the universe was just beginning to take form, all the material which the galaxies were formed from was crowded together. It is also generally stated throughout the literature that a large explosion took place (commonly referred to as the Big Bang), and as such, the material of this infant universe flew apart in all directions, so that at the present time, the distances between all clusters of galaxies are much greater than they were in the past, and will continue to increase at a constant rate.

As you may see by now, if there is one statement that seems clearly remarkable when attempting to describe the universe, it is simply that all galactic clusters are *spreading apart from each other.*

It is advocated by some however, that it is *not* actually the galaxies themselves that are flying off into outer space, but rather, it is *space itself which is actually expanding.*

Now there is simply no question that the galactic clusters are moving away from each other; that is no hypothesis, it is happening indeed!

The argument however, which favors an expanding space, is itself a very ambiguous one, if not a very weak one. I am not sure whether it is even clear to anyone what an expanding space means, for then, *what is there where the space did not yet expand into?* The idea in itself only seems to create additional problems in attempting to *model* the universe, so I will let it drop here for what it is worth.

Back then to the real issue. Now I ask, is it real that we are observing galactic clusters moving away from us at a rate of speed of what appears to be approximately 11 miles per second for every million light-years distant from us! Yes, of course it is real to believe this because it is indeed a fact, an actuality, and a truth, and what's more, it has all been found very objectively by our best astronomers.

This expansion then is actually a true indicator of how fast the galaxies are moving through space, and above all, it is a true indicator of the rate at which we and other galaxies *are accelerating through space.*

Is this expansion not then a Universal Constant of Acceleration—whether it is stated as 8.8 kilometers (5.5 miles) per second per million light years, or 2.8 Angstroms per second per *light* second?

I will answer my own question and say to you, yes! "It is real indeed" to believe that *the expansion rate is strictly the universal acceleration rate*—that is all it can be! It is, for example, the same precise pattern we would find if we could be intelligent microbes riding upon one of those many golf balls I spoke of earlier, falling near the Earth's surface; and except for the difference in direction and the large difference in the ball's rate of acceleration, (32 fps/ps instead of 2.8 Angstroms per second per second for the galaxies) there are no other differences whatsoever. The same basic equation—the same law of nature which applies to a freely falling body on Earth—also appears to apply to the outrushing galaxies.

Now if we will agree that the Hubble expansion rate (on the basis of one light-second) is, in reality, the rate of acceleration of all material bodies in the universe, then it is also real to say that the Earth and everything attached to it are accelerating through space at a constant rate of 2.8 Angstroms per second per second.

So where do we go from here? What else do we know as reality? Well, we know it is real that the velocity of light is constant, that is, constant in that it propagates at one speed when it leaves its source without having to accelerate to get to that speed. From this we can reason that our own velocity through space must then be that which we attribute to light, (since light already possesses this built-in velocity), for how else can light possess such a quality if it were not for the plain and simple fact that it's source is *already moving* at such a phenomenal speed in the first place.

This velocity turns out to be precisely the velocity of our own planet through space if we say the particles our Earth are made of began their journey about 34 billion years ago. For if we then multiply the number of seconds of time in 34 billion years by 2.8 Angstroms we will come up with approximately 300,000 km per second.

In other words, a body such as planet Earth accelerating for approximately 34 billion years at a constant rate of 2.8 Angstroms per second per second, will have a velocity of 300,000 km per second at the end of the 34 billion year period (34 billion years multiplied by 31.6 million seconds equals approximately 1,100 Quadrillion seconds of time). Multiply this product by the 2.8 Angstrom constant rate of acceleration and it equals approximately 3.0×10^{18} Angstroms (300,000 kilometers) per second at our present space-time position relative to the original location in space ($t = 0$) where the particles of our galaxy first began to accelerate from. I will more or less *humorously* call it the vicinity of the "Big Began" since I do not feel the Big Bang theory is the correct one. I believe the universe had instead, a long beginning in the makings of its birth, a very long and very slow beginning, starting out at 2.8 Angstroms per second per second in which it was *not hot* at all, but instead *very cool*, if not cold, by our standards. There were probably no high temperatures which came into existence until millions, or even billions of years have passed in order for velocities to build up sufficiently for particles to agglomerate and eventually heat up into large volumes of hot gaseous bodies.

We should therefore look upon the birth of the universe and its expansion in somewhat the same manner as we look upon all other forms of "Biological Processes" here on Earth; for if there is one thing we surely know of the biological *growth* process, it is that it always begins in a *non-explosive* manner, i.e., with just one or two or a few small particles of matter (never ever exploding instantaneously into being) but rather, very, very gradually developing and growing.

This I believe is really the manner in which the universe itself has grown to what it is at present—*Not* out of a *Big Bang* at all—but more in the spirit of a *very long beginning,* which as stated before, might be more appropriately (if not humorously) ascribed to as the "Big Began!"

8

The Universe in Numbers

The numbers in Table 8–1 may help explain some of the enigmas which presently exist in the field of cosmology, especially with regard to the important parameters of Deceleration, q_o and Hubble's constant, H_o. The value for the Hubble constant, as described in chapter 2 using the acceleration 2.795 Angstroms per second per second, is 8,819 kilometers per second per billion years of time (the first value in column 2-Velocity), or it can be stated as 8.8 kilometers per second per million years (time). Please note that the values in Table 8–1 reflect a linear "velocity vs. TIME" (Fig. 8–1 plot of velocity vs. age) and not the customary "velocity vs. DISTANCE" which is normally associated with the Hubble constant. Fig. 8–2 (plot of velocity vs distance) illustrates the relationship between the velocity and distance values in Table 8–1, and it is *not* a linear relationship.

Age (BY)	Velocity (km/sec)	Galactic Clusters From Time-Zero (BLY's)			Distance Light Years
ι = 0	0,000				00,000,000
1	8,819				14,687,366
2	17,638				58,749,462
3	26,457				132,186,290
4	35,276				234,997,849
5	44,095				367,184,140
6	52,914				528,745,162
7	61,733				719,680,915
8	70,552		1		939,991,399
9	79,371				1,189,676,614
10	88,190				1,468,736,561
11	97,009				1,777,171,239
12	105,828				2,114,980,648
13	114,647				2,482,164,788
14	123,466		3		2,878,723,660
15	132,285				3,304,657,262
16	141,104				3,759,986,596
17	149,923				4,244,648,661
18	158,742				4,758,706,458
19	167,561		5		5,302,138,985
20	176 ,380				5,874,946,244
21	185,199				6,477,128,234
22	194,018		7		7,108,684,956
23	202,837				7,769,616,408
24	211,656				8,459,922,592
25	220,475		9		9,179,603,507
26	229,294				9,923,659,153
27	238,113		11		10,707,089,530
28	246,932				11,514,894,439
29	255,751				12,352,074,470
30	264,570		13		13,218,629,047
31	273,389				14,114,558,352
32	282,208		15		15,039,862,385
33	291,027				15,994,511,499
34	299,846	(Local Group)	17	(MW Galaxy)	16,978,594,646

Table 8–1. The Universe in Numbers assuming a constant acceleration of 2.795 Angstroms per second per second. Note that the distance values, Column 4, are determined by dividing the actual distance traveled in a given time (column 1) by the actual length of our present light year—about 9.5 trillion kilometers.

Age (BY)	Velocity (km/sec)	Galactic Clusters From Time-Zero (BLY's)	Distance Light Years
35	308,665		17,992,022,873
36	317,484	19	19,034,825,831
37	326,303		20,107,003,521
38	335,122	21	21,208,555,942
39	347,941		22,339,483,094
40	352,760	23	23,499,784,977
41	361,579		24,689,461,591
42	370,398	25	25,908,512,937
43	379,217	27	27,156,939,014
44	388,036		28,434,739,822
45	396,855	29	29,741,915,361
46	405,674	31	31,078,465,632
47	414,493		32,483,412,084
48	423,312	33	33,839,690,367
49	432,131	35	35,264,364,831
50	440,950	37	36,718,414,026
51	449,769		38,201,837,953
52	458,588	39	39,714,636,611
53	467,407	41	41,256,810,000
54	476,226	43	42,828,858,121
55	485,045		44,429,280,972
56	492,864	45	46,059,578,555
57	502,683	47	47,719,250,869
58	511,502	49	49,408,297,914
59	520,321	51	51,126,719,690
60	529,140	53	52,874,516,198
61	537,959	55	54,651,687,437
62	546,778	57	56,458,233,407
63	555,597		58,294,154,108
64	564,416	59	60,159,449,541
65	573,235	61	62,054,119,705
66	582,054	63	63,978,164,599
67	590,873	65	65,931,584,226
68	599,692	67	67,914,378,583

Table 8–1. *(Continued)*

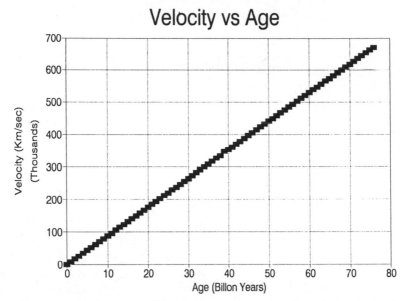

Fig. 8–1. Plot of Velocity vs Age using the values in Table 8–1.

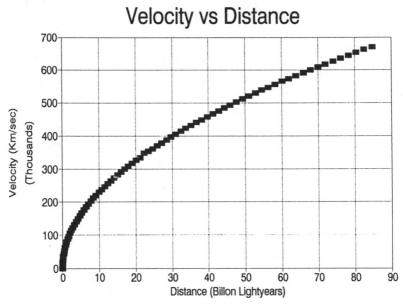

Fig. 8–2. Plot of Velocity vs Distance using values from Table 8–1.

The significant point here is that this constant of 8.819 kilometers per sec per million years refers to *time* and not to *distance,* as in the current use of the Hubble constant. Also, as indicated in chapter 2, the value of 8.819 is approximately one-half the value presently used for the Hubble constant. Another point to note is in reference to our present position in space-time relative to time-zero, $t = 0$, where some workers believe the Center of Mass of the universe originated. It does not appear to be coincidental that the local group of galaxies to which the Milky Way belongs, happens to be at the unique position in the universe where our travel time in years (34 billion) is exactly twice our distance in light years (17 billion) from time-zero. This situation exists because the standard light year, as determined by our present speed of light, 300,000 km/sec, has been used to determine the distance. Therefore, if measured from a point on any galaxy throughout the universe, the age in years will always be twice its distance in light years when using that particular galaxy's value for the speed of light and the size of the light year that particular speed produces. Table 8–2 illustrates this for a galaxy other than the Milky Way.

The values in Table 8–1 indicate that the Milky Way, positioned at 17 billion of our present light years of about 9.5 trillion kilometers each from time-zero, has increased its velocity an average of 17.638 kilometers per second for every million of our present light years of distance it has traveled, or 17.638 kilometers per second for every two million years of time. So that we may state that the Hubble constant really should be equal to 8.819 kilometers per second per million years of *time;* exactly half the present estimated value per million light years. Please note that I did not use light years with my own estimate, but rather I used years because, as related here, the constant applies strictly to the travel time through space by a

Parameters	Galaxy "X"	Galaxy "Y"
Age in years	1.6×10^{10}	5.0×10^{10}
Dist. from t = 0 in earth lightyears	3.8×10^{9}	3.67×10^{10}
Distance from t = 0 in kilometers	3.6×10^{22}	3.49×10^{23}
Light speed in kilometers/second	1.41×10^{5}	4.41×10^{5}
Length of new lightyear in kilometers	4.45×10^{12}	1.39×10^{13}
Distance in new lightyears	8.0×10^{9}	2.5×10^{10}

Table 8–2. Relationship between age in years and distance in lightyears as measured by the speed of light at position of galaxy relative to time-zero.

galactic cluster, and not to *distance* as it is customarily applied to the Hubble constant.

However, also note that in the vicinity of 3 to 5 billion light years in either direction of the Local Group (which includes the Milky Way) you could reach the conclusion that a galaxy's recessional velocity is almost directly proportional to its distance from us just as Hubble discovered (Hubble, Edwin P., 1936, Realm of the nebula: New Haven, Ct., Yale University Press). This is illustrated in Table 8–3 and Figs. 8–3a and 3b.

Thus, from any other galaxy, a hypothetical observer looking out 3–5 bly will, likewise, have good reason to suspect that recessional velocities and distances have a *linear* relationship as the present Hubble Law connotes. However, according to the values in Table 8–1, once greater distances are reached, the apparent linear relationship between distance and velocity is no longer valid, as illustrated in Fig. 8–2. What actually occurs can be seen by comparing values toward time-zero and values for galaxies in the opposite direction.

Also, the values given in Table 8–1 show the following relationship. Even though the velocities of all galaxies

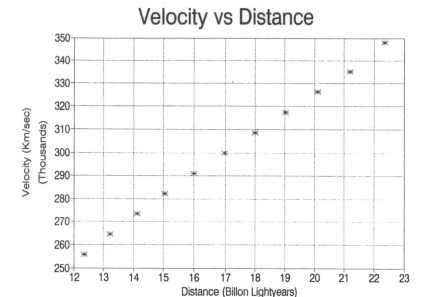

Fig. 8–3a. Velocity vs Distance for galaxies within 5 billion lightyears of the Milky Way. Values plotted here are from Table 8–1.

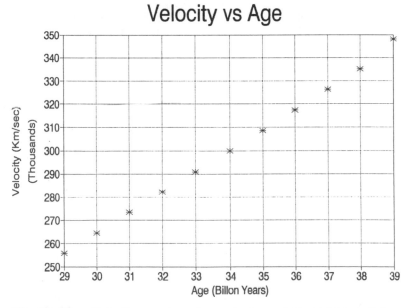

Fig. 8–3b. Velocity vs Age for galaxies within 5 billion lightyears of the Milky Way. Values plotted are from Table 8–1.

Age Billion Years	Velocity km/sec (x 10^5)	Distance (BLY)	Dist. between Galaxies (BLY)
29	2.558	12.352	
			0.86
30	2.646	13.219	
			0.89
31	2.734	14.115	
			0.92
32	2.822	15.040	
			0.95
33	2.910	15.995	
			0.98
34 (MW)	2.998	16.979	
			1.01
35	3.087	17.992	
			1.04
36	3.175	19.035	
			1.07
37	3.263	20.107	
			1.10
38	3.351	21.209	
			1.13
39	3.439	22.339	

Table 8–3. Age, velocity, and distance within 5 billion light-years of our position (MW). These data are from Table 8–1, and are used to create Figures 8–3a and 8–3b.

constantly increase by 8,819 km/sec (5,480 miles per sec) per billion years of time, their recessional velocity per *unit distance* continually *decreases* as they travel farther and farther, and faster and faster, from their original position of time-zero. However, those galaxies that are traveling slower than we are, i.e., that are inward toward time-zero and therefore younger than our Milky Way, will have recessional velocities between them per unit distance to be increasing. Whereas if you are looking in the opposite direction away from time-zero, the change in recessional velocities between any two galaxies will be decreasing. Table 8–4 illustrates these relationships by using some selected, but extreme situations for galaxies that are two billion light years apart. The net result is that as data are gathered for galaxies farther and farther toward t = 0, it

Age (BY)	Distance from time-zero (BLY)	Galaxy vel. km/sec (thousands)	Chg in recessional vel. km/sec per 2 BLY (thousands)	Age diff. (BY)
8.4	1	74	54	6.1
14.5	3	128		
20.5	6	178	28	2.8
23.3	8	206		
34.0	17	300	18	2.0
36.0	19	318		
67.0	66	591	9	1.0
68.0	68	600		

Table 8–4. A comparison of recessional velocities per unit distance relative to each other for galaxies *toward* time-zero and *away* from time-zero.

can lead to the misleading conclusion that the expansion is *accelerating* rather than *decelerating,* as is found in the opposite direction of t=0, when in fact it is absolutely constant *in either direction* in relation to time.

The values in Table 8–1 leads to the conclusion that when we look toward the outer edges of our own material universe and detect large "red shifts" that equate with recessional velocities close to our own speed of light, 300,000 km/sec, it may be that we are detecting galaxies that are 50 to 51 billion of our current light years, and not 17 billion light years as is now believed. For example, the distance in the first column, Column A, depicts the distance between our local group at the 34 billion year epoch and galaxies, if any exist, at the 68 billion year epoch. The recessional velocity of 300,000 km/sec is the difference between our present velocity of 300,000 km/sec (at 34 by age and 17 bly distance) and the velocity of the hypothetical galactic cluster traveling at 600,000 km/sec at its age of 68 by (also the 68 bly position as well). Again, it must be emphasized that at the 68 by epoch the speed of light would be double its present value, hence, the *light year* for those galaxies would be about 19 trillion kilometers instead of the 9.5 trillion kilometers we use for our galaxy.

There are several curious relationships that emerge from this model and from the data in Table 8–1. In our first 34 billion year period we have traveled only 17 billion of our present 9.5 trillion kilometer light years from the position of time-zero, but in another equal period of 34 billion years we will have traveled an additional 51 billion of our present light years and will be then 68 billion light years from time-zero. However, as the speed of light and the length of the light year changes with time, after the second 34 billion years, light is now traveling at 600,000 km/sec and the light year will also double in length. This means that when our galaxy is at this position our age would be 68 billion years old, but, using the new light year length, it would be at 34 billion light years from time-zero. In other words, no matter where we are in the system, our distance, in the light year length for that position and light travel speed, will always be half the age in years.

Depending on where and in which direction we look relative to time-zero, the distances as we currently measure them will be different for the same measured recessional velocity. If we are looking away from the direction of time-zero and see a galaxy with a recessional velocity, relative to our position, equal to 300,000 kilometers/second, its distance from us would be about 51 bly (as we currently measure them). This is because that material has been accelerating exactly *twice* as long as we have and, therefore, has attained *twice* our galaxy's velocity and *twice* the speed at which light travels in our galaxy. In contrast, whenever we penetrate deep in the direction of time-zero, we see material which is younger than our own

Milky Way. If we can peer far enough in that direction, there, also, should be galaxies with recessional velocities of 300,000 kilometers/second between us—the present value at which we see light traveling. However, in this model of the universe, those particular galaxies are actually moving slower than ours since they began their journey through space well after the material found in our galaxy. Because their velocity is slower than ours, they could be only at a maximum distance ($t = 0$) of 17 billion light years (using our present light year length) instead of the 51 bly described above when looking in the opposite direction from time-zero, i.e., the outward portion of our universe.

There are considerable differences, however, about detecting galaxies and other nebulous matter of the universe when we attempt to penetrate inwardly (that is, in the direction of time-zero) as in contrast with looking toward the outer edge. As we are 34 billion years old and 17 billion light years distant from time-zero, we cannot expect to observe images of younger material or galaxies which are actually located much farther back than 8.5 billion light years having an age of no less than 24 billion years since time $t = 0$. This is because of the way light behaves as images are produced. Light from any matter, galaxy, etc., which is less than 24 billion years old has not yet had time to penetrate deep enough through space to reach our own present position of 17 billion light years from time-zero. In other words, from our present space-time position, we should be able to detect the images of galaxies younger than we are which are presently located at a distance of approximately 8.5 billion light years (using the present light year distance of about 9.5 trillion kilometers). These galaxies have an age of no less than 24 billion years and a velocity through space of no less than 212,000 kilometers per second, which is also the speed of light at their position. Any material that is younger than 24 billion years would have a lower velocity through space and therefore, their images, whose light is moving at a lower velocity, would not have time to reach our position.

Given this model of the universe, one in which all particles are experiencing a constant acceleration of about 2.8 Angstroms per second per second, it would follow that in the direction of time-zero there should exist more nebulous matter and maturing galaxies per unit volume of space than that which we would observe in the direction opposite from time-zero.

The other difference in peering inwardly toward $t = 0$ would be that at these distances and ages, much of the slower propagating material of the universe may not have yet grown or formed into mature galaxies of any distinction. In any case, any images of material matter we might possibly detect that is 8.5 billion of our light years distant toward time-zero would have to be of the 2c radiation of that material's *birth* at time-zero (see Figs. 9–9 and 9–10) for examples. Actually, at approximately that point in space (8.5 billion light years from time-zero or half our distance towards time-zero) we should, *hypothetically*, be able to detect the makings of the *first moments* of the birth process in which galaxies would eventually form—the very beginning! Realistically however, we would see nothing so to speak. Of course, closer to us, in the direction of time-zero, we should naturally expect to see more and more, as is presently the case of nebulous matter and maturing galaxies, since they have had more time to develop.

On first examination, it would certainly appear that a deceleration was occurring when we look at material in the *opposite* direction of $t = 0$ and observe older galaxies than our own Milky Way. However, in the model being described here there is a strict and positive acceleration where all the physical laws of nature which apply to freely falling bodies here on Earth, or Mars, or the Moon, or any object with mass, are the same laws which apply to the expanding universe. As indicated earlier, within the universe there appears to be a *universal acceleration* of about 2.8×10^{-10} meter per second per second for all matter, or 8.819 kilometers per second per million light years (at the current length of a light year). As can be seen from the values in Table 8–1, within the vicinity of 3 to 5 billion of our present light years from our present position, the recessional velocities do appear to change at the rate of about 8.8 km/sec/mly (the slope of the line in Fig. 8–3a). These figures are exactly one-half the present value of the Hubble constant of 17.6 kilometers per second per million light years in order to account for a constantly increasing velocity of light.

The table which depicts "The Universe in Numbers" may give a clue as to why it is believed by some astronomers that the expansion process may be slowing. As can be observed from these numbers, it would appear to be erroneous to say there exists a *deceleration* factor in the expansion of the universe. As stated earlier, it is strictly because of *constant acceleration* that, as time goes on, the recessional velocity between any two galaxies becomes less and less *per unit distance* (see Table 8–4). This is the result of each cluster continuing to increase its velocity at a constant rate (2.8 Angstroms per second per second).

Examine the numbers in Table 8–1 again. Notice that after 8.4 billion years a galactic cluster, call it cluster A, at 1 billion light years from time-zero is moving at 74,000 kilometers per second and would have a recessional velocity, relative to time-zero, of the same amount. Another galactic cluster, call it cluster B, that is at the 12 billion year position has a velocity of 101,000 kilometers per second, but it is only 2 billion light years from time-zero, or 1 billion light years from cluster A. However, their recessional velocity per unit distance of 1 billion light years,

Galactic Cluster	Distance from time-zero (BLY)	Velocity km/sec (thousands)	Change in recessional vel. km/sec per BLY (thousands)
t = 0	0	0	
			74
A	1	74	
			27
B	2	101	
C	6	178	
			14
D	7	192	
E	17	300	
			9
F	18	309	
G	67	595	
			5
H	68	600	

Table 8–5. Changes in recessional velocity for galaxies in selected positions relative to time-zero.

27,000 kilometers per second, has decreased dramatically compared with that of cluster A and time-zero, 74,000 km/sec. Table 8–5 illustrates this relationship for galaxies that are only one billion light years apart.

Clusters C and D (Table 8–5) show a similar change. Although they are also 1 billion light years apart, the difference of change in their recessional velocities for this 1 billion light year segment is much less than it was for clusters A and B. If we take a point in space near us, cluster E, and compare it with cluster F, again, only 1 billion light years from us, the change in recessional velocity between the two clusters is now about one-third what it was for clusters A and B. At the position of cluster E, the Milky Way, we have traveled for 34 billion years and are at 17 billion light years from time-zero. A reminder for the reader that these distances in light years are made using our own current measure of 9.5 trillion kilometers for the distance light will travel in one year at 300,000 kilometers per second. However, in the present model, the length of a light year will vary with the changing speed of light, i.e., the light year in "The Universe in Numbers" depends on the velocity in which the galaxies in their time frame move through space. As shown throughout this book, all of this is based upon the postulate that *the velocity of light strictly derives from, and mimics, the velocity of the emitting body.* This in itself may simply explain why nothing can ever exceed the velocity of light.

Once again, getting back to the "mechanics of velocities and recessional velocities of galaxies," look at the values for clusters G and H in Table 8–5. Cluster H, with a speed of 600,000 kilometers per second, is twice as old as our Milky Way (Table 8–1), but is four times as far away from time-zero as the Milky Way, 68 billion light years. At this position the velocity of light for any hypothetical inhabitants of cluster H would, likewise, be 600,000 kilometers per second! Even though cluster H is traveling through space at twice our present speed of light, the recessional velocity difference between it and cluster G, only 1 billion light years closer, is now down to approximately 4,400 kilometers per second. Thus, as indicated earlier in Table 8–4 for a unit distance of 2 billion light years, there is a decrease in rate of change of recessional velocity *per unit* distance.

These then may be some reasons why some astronomers suspect that the expansion of the universe is slowing, i.e., *decelerating.* It only seems to appear there is a slowing because astrophysicists have not generally considered that the universe may actually be expanding under the influence of the same basic Newtonian laws of physics which apply to freely falling bodies on Earth or any other object that has mass.

It would seem, then, from all of this, that in order to *have order* and solve many of the problems which presently confront cosmologists, we may once again have to give Newtonian mechanics the full front seat it enjoyed in the past.

Cutting the Hubble Constant in Half

This hypothesis requires a Universal Acceleration of about 8.8 kilometers/second per million years (per million *light years* in our immediate vicinity), whereas the present estimates for the Hubble constant have a value of around 17.6 kilometers/second per million light years. This necessarily would be a major discrepancy which deserves some comment, if, in fact, all matter is naturally and eternally accelerating at a constant rate, as postulated herein.

The values in Table 8–1 indicate that the light creating the galaxy images on our photographs reached our present space-time position not by traveling at a constant velocity of 300,000 kilometers/second, but rather they arrived here after traveling at various velocities. Table 8–6 illustrates this for observations toward time-zero. When we observe objects that are in the direction opposite time-zero, we are simply running into the faster moving galaxy's zero-velocity (0c) images. (Note: Chapter 9 discusses in detail the 0c through 2c velocities of light.) However, as we only have our own light velocity (1c) to use as a reference, 300,000 kilometers/second, we assume the images have traveled to us with *that* velocity, namely 300,000 kilometers/second. With the hypothesis presented here, it is we who have actually traveled and run into the images of galaxies which are themselves older than our own. Because the velocity of light is a vector quantity, images created in the direction of time-zero behave differently

An Observer:					
on a galaxy with an age of: (BY)	is at a dist. fro t=0 of: (BLY)	moves with that galaxy at a velocity of: (M km/sec)	can observe in direction of t=0, birth of another galaxy with an age of: (BY)	whose actual distance is half its own distance from t=0 or: (BLY)	moving at a velocity of: (M km/sec)
12	2	106	9	1	79
17	4	150	12	2	106
21	6	185	15	3	132
24	8	211	17	4	150
26	10	228	19	5	167
29	12	256	21	6	185
31	14	274	22	7	193
34 (MW)	17	299	24	8	211
37	20	235	26	10	228
48	34	423	34	17	299
68	68	598	48	34	423

Table 8–6. The numbers above are approximate and are taken from The Universe in Numbers (Table 8–1). The values in Table 8–6 reflect the greatest distance an observer on any galaxy can theoretically detect any material substance in the direction of t = 0 (our inner Universe); material which is always younger than the observer's galaxy. See for example, our own position (MW) at 34 billion years in time. The situation described in this table does not apply when an observer is looking toward the "outer edge" of the Universe. In that case, the observer can not observe any hypothetical births of galaxies, but can observe galaxies as they were when they were the same age and distance from t = 0 as the observer's present galaxy, *even though they have different ages* (all older than the observer's galaxy and are physically at various distances from the observer).

than images created on the side away from time-zero. The light forming images on the side toward time-zero will have a net velocity of zero; we are moving in one direction at 300,000 kilometers per second and the light of the image is moving at the same speed, but in the *opposite* direction. Hence, the net speed will be zero and the image will not be moving relative to time-zero. The light in an image created on the side away from time-zero—in the direction we are traveling and accelerating—will not only have the 300,000 kilometers per second (the present speed of light), but it will, also, gain the velocity of the Earth which is traveling in the same direction at 300,000 kilometers per second, or 2c, where c = the present speed of light. A useful analogy would be to think of the images like the exhaust gases from a jet engine in flight. The gases exiting the combustion chamber at the same velocity, but in the opposite direction as the combustion chamber is moving, then the net velocity relative to the ground is zero. If the aircraft is shooting a cannon or a missile, then the missile has not only its forward velocity, but that of the aircraft as well. Therefore, the manner in which we presently determine how light propagates should be revised to conform to the "Principle of the Addition of Velocities" which, as noted previously, will be expounded upon in more detail in the following chapter.

As a consequence of the ideas presented here, the distance scale would have to be doubled in order that the calculation of the Hubble constant (estimated to be about 57.5 kilometers/second per megaparsec or about 17.6 kilometers/sec per million light years) will reflect a more accurate and realistic recessional velocity of exactly one-half the present estimate, or about 28.7 kilometers/sec per megaparsec (8.8 kilometers/second per million light years or 2.8 Angstroms per second per lightsecond.)

When first examined, the numbers in Table 8–1 and 8–6 would appear to produce a value for the Hubble constant which is exactly twice the Universal Constant of Acceleration proposed herein. However, the Hubble constant applies to light years as a *distance* scale, whereas the estimates made with the Universal Constant of Acceleration applies only to years as a *time* scale or *age*. However, as demonstrated in Figs. 8–3a and 8–3b, within 3 to 5 billion light years in any direction of the Milky Way, both values can appear to be associated with either a time scale *or* a distance scale. But for greater distances, the Universal Constant of Acceleration must be associated only upon a *time* scale and *not* a distance scale, i.e., years and *not* light years. Thus, the discrepancy between a time scale and a distance scale will not make much difference until we reach distances greater than 5 or 6 billion light years from our own position—the position of our own local cluster of galaxies which contains the Milky Way.

One way of describing (on a per light second basis) why the Universal Constant of Acceleration is half the present estimate of H_o, is that the average velocity of light must be taken to be 150,000 km/sec instead of its present 300,000 km/sec; this requires that light NOT have a constant velocity over time.

Using this approach then, when astronomers measure the recessional velocity between us and another galaxy, the value of H_o should be 17.6 kilometers/second per TWO (2) million light years, and then from this the distance can be determined. Since the recessional velocities are determined by assuming a constant velocity of light of 300,000 kilometers/second, this forces the value of H_o to be 17.6 kilometers/sec per million light years. However, if the velocity of light is, in fact, *not constant,* but has been increasing steadily since time-zero, then the *average* velocity should be used in the calculations instead of the present value. When that is considered, the average travel time will be slower by a factor of one-half. Hence the value of H_o would be, as suggested above, per 2 million light years. If that is the case, then one-half H_o turns out to be the estimates based on the present hypothesis, namely 8.8 kilometers/second per million light years or, 2.795 Angstroms per second per lightsecond. Once again, however, for the sake of accuracy, these values would be in velocity versus *time,* and not in velocity versus *distance.*

9

On the Way Light Propagates

During the late 1880's, Albert A. Michelson and Edward W. Morley were having a difficult time attempting to prove that light propagated in strict accordance with the "Principal of the Addition of Velocities," such as is the case, for example, with acoustical radiation.

Using the Earth's orbital velocity around the Sun, Michelson and Morley attempted to prove they could detect a difference in the velocity of light by directing two beams of light at a 90 degree angle from each other. Since the Earth orbits the sun at about 18 mps (this chapter will utilize mostly British units), they had naturally believed that a difference in light's velocity should show up when they compared one beam's velocity from the other. They were, however, never able to prove such a variation existed.

Sixteen years later, in 1905, Albert Einstein, in his Special Theory of Relativity, declared that *the speed of light was an Absolute and Unvarying Constant throughout all space and time*—even to the extent that if one could move alongside a light beam at 90% its speed, they would still measure the beam to be moving at 186,282 mps.

The sketches in this chapter outline what I believe is more properly happening when light propagates; and why Michelson and Morley were not able to measure the difference in variation of light's velocity. In other words, beside the fact that light constantly increases by 2.8 Angstroms per second per second, it also varies in its propagating speed, from *zero* velocity 0c to 372,000 miles per second 2c, as Figs. 9–1 to 9–11 will purport to show. What follows then will be a sketch of each figure, and a brief description. Note that the Earth and Moon, *for all practical purposes,* are depicted as being exactly one (1) lightsecond (186,282 miles) apart, even though they are, in reality, about 240,000 miles apart.

In Fig. 9–1, a theoretical observer on another galaxy located at Milky Way image position 2c, hypothetically sees Milky Way's image of its birth. At that moment, Milky Way's birth image has just arrived at position 2c as observer galaxy (located 34 bly's from t=0) is just at that moment in time moving away from 2c image of Milky Way at 265,000 miles per second. Therefore, even though in reality, the Milky Way itself is located at 1c (17 bly's from t=0) traveling at 186,282 mps, *its birth* image is hypothetically observed by observer at 2c position as having a recessional velocity between them of 265,000 mps, and is perceived to be at a distance from observer galaxy of twice its actual distance, since observer at 2c is only seeing *at that moment in space-time* the Milky Way as it was at birth, that is, at its 0c space-time position at t=0. So in effect, when observer at 2c sees the image of the Milky Way as it was at birth, the Milky Way itself is actually only half that distance away in a much more developed stage which observer at 2c cannot yet see.

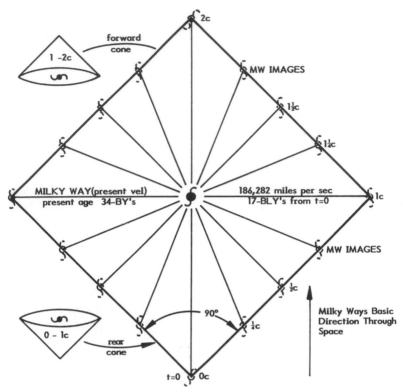

Fig. 9–1. The Milky Way's Two Light Cones—From (0c) Zero Miles Per Second to (2c) 372,564 Miles Per Second.

According to Newton's Law of Motion, the amount of velocity which can be imparted to a body will depend upon the body's mass and the force applied to it; and that any velocity once imparted will be rigidly maintained unless other forces act upon it, i.e., when no force is exerted on a body it remains at rest or continues on in a straight path at constant speed, whichever is the case. Naturally then, the more force that is applied to a body, the faster it will move; and of course, the less massive a body is, the faster that same force will propel it.

In empty space then, where there is normally presumed to be no frictional effects acting upon a body, an object set in motion would simply continue in motion in a straight line with constant speed unless some other force acted to change its direction and or speed. How fast the object travels will depend strictly upon its mass and the amount of force imparted to it. This is known as the Law of Inertia, *Newton's First Law of Motion*, which had first been proposed by Galileo before him.

Let us go on then to a *Photon of Light* and apply this principle of the law of motion to a light beam. Light beams are made up of photons, and photons are presumed to be massless. Therefore, if photons have no mass, and the source from which the photons issue is already in motion—such source moving at a speed of 186,282 mps—what will be the photon's velocity? The answer to this should be evident, for if photons are massless they should propagate away from the source in its *forward* direction of travel at the precise speed of the moving source which they (the photons) are released from. So that relative to a point in space, the photons will actually travel a total distance of 372,564 miles in one second. In other words, the photons are being impelled out ahead of the source at 1v (186,282 mps) while at the same time they are being *carried* along with the source at another 1v for a total velocity, relative to a point in space, of 372,564 mps, or 2v (see Fig. 9–2, Propagation of Light at 2c). This is in strict accordance with the Principle of the Addition of Velocities. However, it will appear to all observers (who themselves do not realize it is the source itself and they who are actually moving at 1v) that the photons *on their own* traveled 186,282 mps, or 1c.

If released from the rear of a moving platform (of which platform we will refer to here as the Earth) the photons will simply stay put, *dropping off in the Earth's tracks,* just as do the *contrails of hot gases* from jet planes in

flight. In this case however, even though the actual velocity of the photons (aimed, or otherwise pointed in the
rearward direction) is zero 0c, it will still appear to all
observers that the photons had shot away from the Earth
at 186,282 mps 1c since the observers themselves are actually moving at that speed along with the Earth, but in
fact do not realize that they (the photons) are not moving
at all. (See Fig. 9–3, Propagation of Light at zero velocity 0c.

If the photons from the light source were released
from the side of the moving platform, that is, directed at
a 90 degree angle to the platform's forward motion (see
Fig. 9–4 propagation of light at 1c) they would then
propagate out at the precise velocity of the body they are
released from—186,282 mps—a speed of 1v of the moving platform, or once again, 1c as refers to the speed of
light.

This then is why light always appears to propagate at
one speed no matter how it is measured. It is because we
have not been aware of the most significant factor, and
that is—that it is we, the Earth, the stars, the galaxy, that
are actually moving through space at the velocity we attribute to light. And so, no matter how, or from which direction we measure the velocity of light, it will always appear to us to propagate at the one unvarying speed of
186,282 miles per second—at least for this moment in
space-time—since for each second of time which goes by,
our own basic velocity actually increases by approximately
2.8 Angstroms; consequently, so too will the velocity of
light follow suit and increase likewise.

This, most apparently then, is the sole reason why Michelson and Morley were not able to detect a variation in
the speed of light relative to the Earth's velocity in its orbit around the Sun. It was simply because the propagation of light itself exists solely as a result of the fact that
the body in which it is emitted from, is itself in motion

moving through space, and that it is this precise velocity of the emitting body which determines the precise velocity of light as we perceive it to be. Therefore, since the propagation of light strictly depends upon the real and actual velocity of the body it is emitted from, it then stands to reason that light's velocity, *relative to space,* should in reality, vary in every conceivable direction— from as much as twice 2c the emitting body's real velocity, all the way down the scale to zero velocity, 0c.

However, as was the case with Michelson and Morley's experiments, there is presently *no known* way one can differentiate between the varying velocities of light, since relative to the observer, light will always appear to have but one velocity; and that is the velocity in which the observer moves through space, no matter which direction the observer looks in, or how it is measured. For example, Fig. 9–1 shows a schematic of the propagation of light from our own galaxy, the Milky Way. This can represent any galaxy, star, planet, or for that matter any mode of electromagnetic radiation located anywhere in the universe. It is designed to depict only the general picture of how light propagates at speeds from zero-velocity 0c to twice the velocity 2c of the emitting body's real velocity, 1v.

Starting with the Milky Way's image, both in the direction of 0c, and at 0c itself, what happens here is that as the Milky Way moves ahead, its image just simply stays in place, *drops off from the rear of galaxy,* just as the contrails of hot gases issue or drop off from the rears of jet engines of planes in flight. In that direction, the actual velocity of light "relative to space" is zero 0c. However, since all observers on our galaxy are moving along with the galaxy at 186,282 miles per second but do not in fact realize it, they will naturally determine the velocity of any light beam they point in the 0c direction to be 186,282 mps (see Fig. 9–3 Earth-Moon schematic).

Going now strictly to the Milky Way's 2c direction of its image in Fig. 9–1, what happens here is that the image is propelled out ahead due to the Milky Ways forward momentum. It is propelled at the precise speed the Milky Way is traveling (Newton's 2nd law of motion) so that whatever distance the Milky Way has actually traveled from t = 0, its earliest *at birth image* has also traveled out ahead of it an additional and equal distance, so that at that point in space the velocity of light, relative to t = 0, is 372,546 miles per second, 2c. Observers however on another galaxy located at 2c would measure *their* recessional velocity with the Milky Ways birth image to be 265,000 miles per second, since at that space-time position an observer at 2c is actually traveling at 265,000 mps. This observer will measure his velocity of light to be 265,000 mps for the same reason we measure ours to be 186,282 mps. See for example, Fig. 9–2, Earth-Moon schematic.

What happens also at galaxy image 1c in Fig. 9–1 is simply that the Milky Way's forward momentum causes its image to be propelled out at right angles at the same speed that it travels. Because the galaxy moves forward it drags its image right along with it, and although the image appears to travel the long path of the *hypotenuse*, it actually travels the shorter distance of the lines at right angles, hence, the velocity of light in this case is precisely 186,282 mps 1c, both for an observer, and relative to space. Further details of this can be seen in Fig. 9–4, Earth-Moon schematic. In addition, Figs. 9–5 and 9–6 detail the light rays at velocities of ½c and ¼c respectively.

In Fig. 9–2 the Moon and Earth are moving together through space—the Earth directly behind the Moon—at a velocity of 186,282 mps. For the sake of convenience they are positioned precisely one light second (186,282 miles) apart from each other instead of their actual average distance of 234,000 miles.

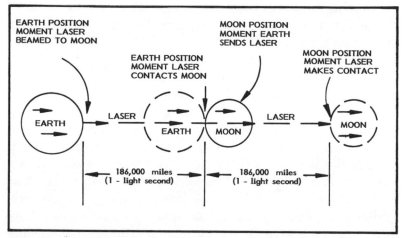

Fig. 9–2. Propagation of Light at (2c) 372,564 Miles Per Second.

How does the laser beam from Earth arrive at the Moon? What is the basic mechanism of propagation? It is simply that as the Earth moves forward at 186,282 mps, the beam, having no mass to speak of, is automatically propelled out ahead of the Earth at precisely the same speed in which the Earth moves through space (this in accordance with Newton's second law of motion). By the time one second has passed, when the Earth has advanced 186,282 miles to the position the Moon was originally at when the beam was initially directed to the Moon, the Moon will also have advanced thru space 186,282 miles.

What actually happens then, is that the Earth carries, or pushes the beam along with it at the velocity it is traveling, 186,282 mps; and in addition, it also propelled the beam out ahead of it at *another* 186,282 mps (according to Newton's second law of motion) so that in effect, the beam traveled—relative to the original position of Earth when it released the beam—372,564 miles.

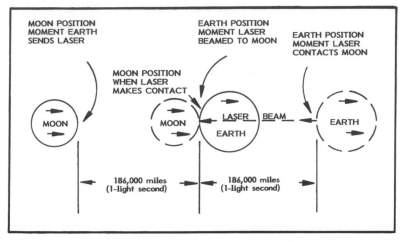

Fig. 9–3.   Propagation of Light at Zero Velocity (0c).

In Fig. 9–3, both Moon and Earth are moving together through space, the Moon directly behind the Earth at a velocity of 186,282 mps. For the sake of convenience they are positioned precisely one light second (186,282 miles) apart from each other instead of their actual average distance of 234,000 miles.

How does the laser beam get to the Moon from the Earth? What is the actual mechanism of propagation? It is simply that as the Earth moves in its forward thrust through space, the beam that is aimed at the Moon from the Earth's rear position just simply stays put, i.e., it drops off in its tracks as the Earth pulls away from the beam—once again—just as the contrails of hot gases drop off in the tracks of jet planes in flight. So that in effect, the Moon runs head on into the laser beam at the speed with which the Moon moves through space, 186,282 mps.

Here again however, observers on Moon and Earth do not realize it is they that are moving through space at 186,282 mps, and therefore, both simply believe the laser beam—on its own—propagated from Earth to Moon at

186,282 mps, when in reality the beam actually traversed space at a speed of 372,564 mps since it was pushed, or otherwise carried along with the Earth for 186,282 miles in that one second journey, while at the same moment, it was thrown outwards or otherwise propelled ahead of the Earth for an additional distance of 186,282 miles.

In Fig. 9–4 the Moon and Earth are moving together through space side by side at a velocity of 186,282 mps. For the sake of convenience they are positioned precisely one light second (186,282 miles) apart from each other. How does the laser beam from the Earth arrive at the Moon? Not so simple in this case. As the Earth and Moon move forward through space the beam from Earth is aimed at the Moon from a 90 degree angle in relation to Earth's forward thrust through space. As the Earth advances through space along with the Moon, side by side, each at 186,282 mps, the beam from Earth is naturally carried along with it. However, since it is directed from a 90 degree angle (in relation to its forward thrust through space) towards the Moon, the speeding Earth also hurls or shoots the beam out at precisely the same velocity the Earth moves (this again according to Newton's second law of motion). The Earth's forward thrust through space of 186,282 mps is precisely applied to the massless laser beam—aimed from a right angle of its forward thrust direction—so that there is a misleading appearance of the beam actually traveling the longer length, *geometrically known as the hypotenuse in this triangular sketch,* when in fact it is continuously propagated from Earth at right angles—relative to their forward motion through space— for a total distance of 186,282 miles. Here again, observers on both Earth and Moon simply do not realize it is they who are actually moving through space at 186,282 mps; and so, they naturally believe the beam propagated in one single straight line at 186,282 mps, instead of the more broad and complicated path it took in reality.

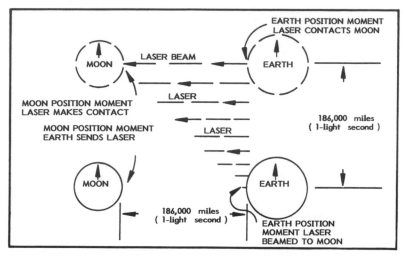

Fig. 9–4. Propagation of Light at (1c) 186,282 Miles Per Second.

In Fig. 9–5, the Moon is one-light second away at a 45 degree angle behind Earth in their forward thrust through space. Since Earth releases laser beams at a 45 degree angle from its rear quarter position, the beams' actual velocity relative to its space vector will thereby be half the speed the Earth itself travels, or 93,141 mps. The other half of the beams apparent velocity of 186,282 mps is due to the effect of the Earth pulling away from it. Thus, as the beam is thrust towards the Moon, spreading at half the speed of Earth's velocity through space, it is also deposited in place as the Earth races away from it and the Moon rushes towards it; so that it will then appear to observers on Moon and Earth that the beam simply shot out from the Earth at 186,282 mps in one single line of propagation. Neither realize they have themselves each advanced through space 186,282 mps while the beam (in one second of time) was finding its way to the Moon at only half the actual velocity which they them-

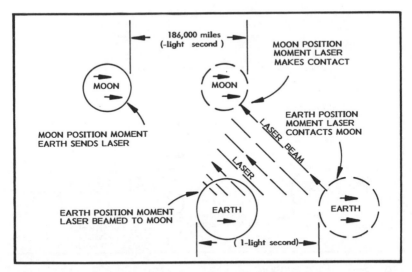

Fig. 9–5. Propagation of Light at (½c) 93,141 Miles Per Second.

selves are moving—half the apparent velocity of light. It is, in other words, the propagation of light as a combination of the action which Figs. 9–3 and 9–4 depict.

In Fig. 9–6, for the sake of convenience, the Moon is one light second away, but only a 22½ degree angle behind the Earth in their basic thrust through space. Since Earth releases the laser beam at a 22½ degree angle from its rear quarter position, the beam's actual velocity—relative to space—will thereby be only ¼ of the speed the Earth itself travels, or 46,570 mps. The other ¾ of the beam's apparent velocity of 186,282 mps is due to its staying put, *just simply dropping off in its tracks and spreading,* as the Earth moves away from beam and the Moon rushes towards it. So that as the beam spreads at ¼ the speed of Earth's velocity through space, it is also deposited in place as the Earth pulls away from it. Consequently, it will then appear to observers on Moon and Earth that the beam

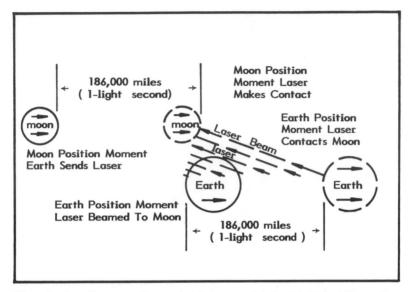

Fig. 9–6. Propagation of Light at (¼c) 46,570 Miles Per Second.

simply propagated from Earth at 186,282 mps in one single line of travel.

They will not realize they have each advanced through space 186,282 miles as the beam, in one second of time, was finding its way to the Moon at only ¼ the actual velocity which they themselves are moving, or ¼ the apparent velocity of light.

The schematics in Fig. 9–7 depicts the ages and distances traversed by three galaxies, X, Y, Z, and their *double light cones*. The distance scale is based upon our own present light year of 5.9 trillion miles. The time scale is based upon our own length of year of 31.6 million seconds.

Notice that the distance in light years is not equal to the ages in years. That is simply because they are based

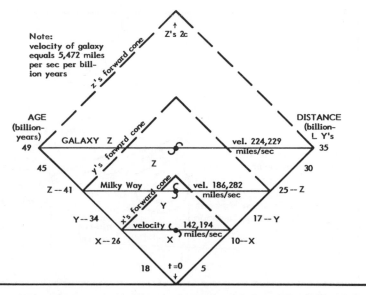

Fig. 9–7. Three Galaxies and Their Lifetime Double Light Cones.

upon our own space time standards in this schematic. Also, it is no coincidence that galaxy Y, the Milky Way, is 17 billion light years in distance from time-zero and exactly twice that number, 34 billion years in age. As stated previously, this 2 to 1 ratio is not unique to us. Again, that is simply because our present light year scale is utilized in the calculations (See for example Fig. 8–1, The Universe In Numbers).

The three double light cones in Fig. 9–7 are due to the fact that a galaxy's light—in its forward direction through space—is propagated at velocities between one and two times the actual velocity of the galaxy from which the light issues. For example, galaxy Z's image propagated at twice its velocity, 2c, or 448,704 mps as indicated at the extreme upper point of its broken line cone, and has propagated at one unit of its velocity, 1c, or 224,352 mps to the extreme right and left of its actual position. However, at the extreme bottom portion of its solid line (rearward) cone, its velocity of light is zero, 0c.

The velocities of light in between these examples can simply be determined by the process of extrapolation. Any light source, whether its a galaxy, a star, a planet, or for that matter any other emitting source of electromagnetic radiation, will naturally create a double cone of its signal, image, etc. and will propagate through space at velocities ranging between zero 0c and twice (2c) the actual velocity of the source itself. However, the apparent velocity of 1c (as in contrast to the real velocities 0c to 2c) will always seem to be at 1c no matter which direction it is measured from, and no matter how it is measured, just as the famous Michelson-Morley experiment proved to be when they themselves attempted to detect the variations in the velocity of light.

In Fig. 9–8, galaxy (A) represents the Milky Way at a distance of 17 billion light years from time-zero after 34 billion years of constant acceleration at 2.8×10^{-10} meter per second per second (approximately 2.8 Angstroms per second per second). Its image (light) is propagated in all directions—from zero velocity, 0c, to twice its actual velocity, 2c. It is this variation of the velocity of light (from zero to twice that of the actual velocity of the emitting body) which creates the double light cone. This same picture applies to galaxies B, C, and D. The only differences are that B is 14 billion of our present light years from time-zero after 31 billion years of travel at a constant acceleration of 2.8 Angstroms per second per second, while C is 11 billion light years from time-zero after 27 billion light years of travel at the universal constant of acceleration rate, as in B above; and D likewise, is 9 billion light years from time-zero after 25 billion years of travel.

Their present velocities through space appear in the schematic of Fig. 9–8. It is not simply a coincidence that

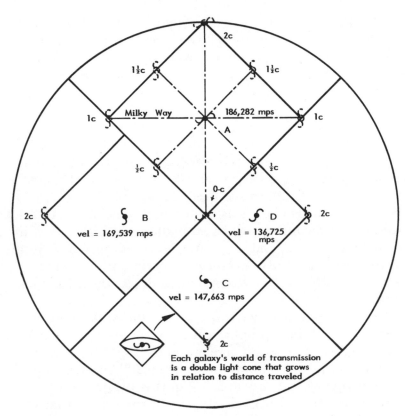

Fig. 9–8.   The Universe—Its Galaxies And Their Light Cones.

our own galaxy's age of 34 billion years is exactly double in "years of *time*" as its 17 billion "light years of *distance*." As can be seen from Fig. 8–1 (The Universe In Numbers) this 2 to 1 ratio is no exception but is rather the *rule* for all galaxies throughout the universe. This is simply because different velocities will naturally result in different light year distance scales; and so, based upon any galaxy's present light year, that particular galaxy will always be

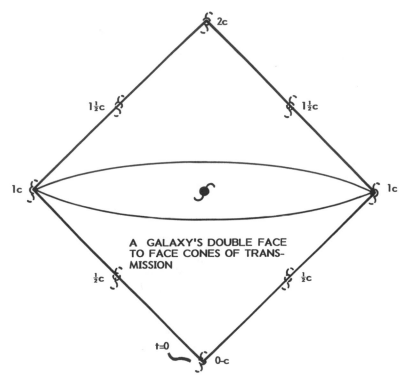

Fig. 9–9.   The Double Face To Face Cones of Transmission.

twice its age in years that it is in distance of its own light year from time-zero.

In Fig. 9–9, each and every material body in the universe, whether a particle or a star; transmits its image, signal, etc. in the form of a double *face* to *face* cone. These double light cones are the result of the body's basic forward thrust through space. The distances they will extend to will be equal to the actual distance the body traveled. Naturally though, a body's temperature, and or signal strength, will always affect the extent of its observable transmissions, as for example, with very bright Quasars—

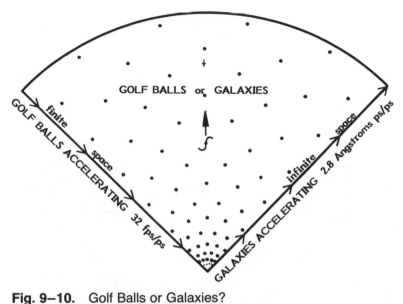

Fig. 9–10.   Golf Balls or Galaxies?

for although their light cones naturally exist by virtue of
their travel time from $t=0$, their extremely bright light
deems them to be more readily detectable.

In Fig. 9–10, Uniform Expansion can have but one
interpretation—and that is—Acceleration! Otherwise, it
would be impossible to say that the universe is expanding
in the Hubble sense if we do not mean to imply that the
galaxies are *accelerating* at some constant rate.

The picture of Fig. 9–10 is no different for galaxies
flying apart from each other, i.e. *accelerating,* than it would
be for a quantity of golf balls *falling inwardly* towards each
other to Earth under the influence of Earth's gravitational field. However, the golf balls would initially start
out from the top of the picture in Fig. 9–10 and accelerate towards the center of the Earth at a constant rate of
32 ft. per second per second.

The galaxies motions differ from the golf balls in that
the galaxies accelerate in the opposite direction towards

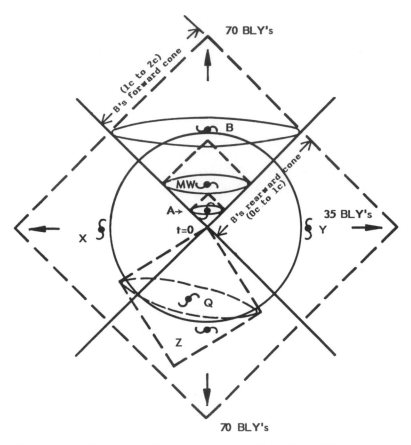

Fig. 9–11. Who Can Observe Who In This Slice Of The Universe?

the top of the cone in a "spreading fashion" at only a minuscule rate of acceleration of one-ninety-millionth of an inch (2.8 Angstroms) per second per second. Otherwise, the laws of nature for both systems are identically the same; that is, the doubling of time at any constant rate of acceleration always equals a quadrupling of distance traveled. And that is so, simply because in any system of constant and uniform acceleration, a doubling of time means also, a doubling of the velocity. That of course is why distance quadruples at twice the time.

Fig. 9–11 shows a sample of the Universe in which MW represents the Milky Way moving at its present velocity of 186,282 mps. MW is 17 billion light years from time-zero after accelerating through space for 34 billion years at a constant rate of 2.8×10^{-10} meter per second per second. Its forward light cone extends an additional 17 billion light years so that its image at birth is detectable at 34 billion light years from time-zero at a point in space just short of galaxy B. The outer edge of this sampling of the material universe (galaxies B, X, Y, and Z) extend out to 35 billion light years from $t = 0$, while each of their forward light cones extend out to 70 billion light years from time-zero.

In this scenario, galaxy (A) can hypothetically observe galaxies MW and B providing that its observers possess sufficient detection power. MW can observe B but not (A) since A's forward light cone does not extend to MW. B cannot yet observe MW's birth since MW's forward light cone comes just short of reaching B; and of course, B cannot observe (A). Z can observe Q and Q can observe Z. Neither X nor Q receive each others light, for as Fig. 9–11 depicts, although each of their light cones extend into the others light cone, neither light cone extends out to either galaxy.

10

Are Atoms Really Empty?

According to Classical Physics, ATOMS are said to consist of a *trillion* parts empty space to *one* part solid matter. A trillion to one is really being conservative—the actual ratio is even greater. Some sources put it as high as a quadrillion to one, but I'll work with the trillion to one ratio.

If you should ever wonder: "Just How Empty Are Atoms" you would be correct—according to the laws of classical physics—to believe that ATOMS are *so empty*, that if all their protons, neutrons, and electrons from say 1.6 million tons of solid steel were hypothetically stacked together touching one another, they would require a volume of space of not more than that which a *single grain of RICE* occupies in the normal sense. That's right! Each and every solid particle—every proton, neutron, and electron—which 1.6 million tons of steel is comprised of, could in accordance with the laws of classical physics—fit in a container no larger than one required to hold a Single Grain of Rice.

In even more graphic terms, think of a 10 foot high solid mass of steel having the width of a 10 foot wide highway lane, and a length which extends for TWELVE (12) MILES down the highway. This is the actual volume of space which 1.6 million tons of solid steel would require; yet for *all* that steel, each and every bit of solid matter which it is comprised of (every proton, neutron and electron) would—if *stacked together touching one another*—displace a volume of space not more than that of One Grain of Rice.

What would the Earth itself boil down to if it were hypothetically voided of all the empty space which each atom is supposed to consist of? In this case, if only the solid stuff of the Earth's atoms (it's protons, neutrons, electrons) could be formed into a cube, it would consist of no more than a cube three football fields long by three football fields wide by three football fields high. That of course would amount to a three hundred yard (900 feet) cube. But, of course, to depict a picture of just how empty the atom is conceived to be, I believe the first picture of the 1.6 million tons of steel, the 12 miles of highway, and the single grain of rice tells the story best.

These graphic examples provide for an excellent *mental construct* of: not just "how empty" atoms are believed to be; but also, it provides an excellent manner in which to envision just "how incredibly small," protons, neutrons, and electrons are believed to be. For myself, it makes me ask this question: Do physicists realize just how incredibly *small* they made out the protons to be; and just how incredibly *empty* they made the atoms to be??

This is the kind of stuff inquisitive students may tend to challenge. For example, one interesting question might be: "If atoms are so empty, then how can TRANSPARENT mediums such as GLASS, AIR, or WATER *slow* the massless photons of *light* traveling thru them at the phenomenal speed of 186,000 miles (300,000 km) per second?"

What then is the *classical* structure of the Atom like? In other words, what kind of a mental construct can one make of it?

The best picture to describe the atom is one in which it is composed of a central massive core, the *nucleus*, which is comprised of all the atom's protons and neutrons. This leaves us with *all* of the atom's *electrons* (there is one electron for every proton) circling around outside the nucleus (each at a rate of billions of revolutions per millionth of a second) in which the orbital radius of some of the more massive atom's electrons extend as far out to a distance from the central nucleus, of at least the length of 10,000 times the nuclear radius, or by some estimates, 100,000 times the nuclear radius.

That is the classical picture of the average atom of matter. From that description, one can readily envision just why the atom is believed to be so empty. Since the orbiting electrons are said to extend out to a distance from the atom's nucleus as far as 10,000 times or more of the nuclear radius; and since the point like electrons (10^{-18}m diameter) are believed to be so small as to take up no space to speak of, then the total volume of empty space in an atom is easily calculable as the *cube* of 10,000, which is of course *one-trillion*. This then is how the phenomenal ratio of one-trillion parts empty space to one part solid matter is derived.

Physicists investigate the atom's particles primarily with the use of Particle Accelerators. Particle accelerators are used to accelerate protons, for example, to higher and higher velocities in order to collide them with sufficient energy to give physicists a picture which describes as much as possible the proton's characteristics and physical parameters. They investigate and study many other types

of particles including the electrons and the recently discovered *Quarks,* the particle which physicists now believe
is the basic constituent of all matter where, prior to the
Quark's discovery, protons and neutrons were believed to
be the basic constituents of all matter. However, although
it is believed that protons and neutrons are now comprised of three quarks each, physicists have yet to actually
isolate a *free* quark. For this reason, some physicists do not
accept the quark theory in spite of the fact that it has
popular support throughout the scientific community.

A Super Conducting Super Collider is presently under construction in the state of Texas. This will be the
largest (54 miles in circumference) of all the Particle Accelerators in the world. It is with this super collider in
which physicists are seeking to find more answers, not
just of the Quark, but of several others such as the infamous Higgs Boson, a super heavy particle which is sus-
pected to be even more massive than the W and Z particles which, themselves, are believed to be 100 times as
massive as the proton, with the Z particle being even
heavier than the W particle by about 15 percent.

So whatever the details of the atom's structure may
actually be, it is inconceivable that all matter could actually be comprised of approximately 10^{12} to 10^{15} parts
empty space to one part solid matter; and inconceivable
that protons and neutrons are so small that 1.6 million
tons of steel would hypothetically reduce to the volume
of one grain of rice.

The graphic examples above then, are striking repre-
sentations of the classical picture one would have to accept in order to hold to a belief that *all* electrons normally
orbit *outside* the nucleus.

Instead of having atoms which are full of empty
space, wherein the key factor contributing to that picture
is the electrons orbiting outside the nucleus, i.e., up to
10,000 times the distance of a nuclear radius, it would

seem to be more realistic to believe that "each electron" normally orbits an individual proton (as is the actual case with *hydrogen* atoms) while neutrons should play the major role of stabilizing the atom by keeping protons separated from protons in order that their super strong *repulsive* force which they—the positively charged protons—would exert upon each other, will not have the devastating effects which could be expected of such a force. In other words, without the neutrons normally keeping protons in each nucleus separated, the repulsive force which protons have upon each other would likely *"blow the atom apart."* However, it is most likely this *"repulsive force of protons"* which is the key in determining the density of matter which will be further explained below.

As for this proposed atomic model of electrons *orbiting individual protons,* electrons would still be found outside the nucleus, as is commonly the case; however, this would be due to those electrons which are orbiting the atom's *outer* protons; and there are most likely as many, or more of these situations, than there would likely be of electrons orbiting *inner* protons.

This *"not so empty atom"* would preferably result in the structure of atoms being more like that which is represented in Fig. 10–1a. So that instead of the helium atom being constructed as in Fig. (b), the classical method, it would be constructed more like that of (a). This type of "atomic construction" would be representative of all atoms, no matter how many protons and neutrons they consist of. In simple terms, protons of all atoms are always separated by neutrons.

In the schematic of Fig. 10–1a, protons would not normally touch each other, which, if they did, would likely cause the atom to be unstable and, consequently, annihilate. Instead, electrons would have open space to orbit freely around each inner or outer proton, *there being four different paths each inner electron would be free to take.*

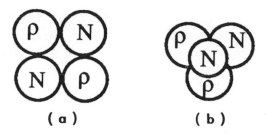

(a)   (b)

Fig. 10–1.   Restacking the Nucleons

The nucleus would therefore be the *complete* atom; and the total empty space in an atom should amount to not more than possibly *two parts space to one part solid.*

A less significant argument in support of this "not so empty atom" might be along these lines: Since there is one electron in every atom for every proton (both having opposite but equal charge), why then should the electrons *desert* this most favorable position for one which makes no *biological sense* at all, i.e., offspring electrons should stick close to "parent" proton—just as mother nature would have it; or, just as is the case with the hydrogen atom—the most abundant atom in the universe.

Finally, it can be argued that this type of structure for atoms may not be so far fetched after all. Why? Because for one thing, *no one has yet isolated a free Quark!* And if this proposed model of the atom's structure, as shown in Fig. 10–1a, could possibly be correct, it would then have to be that Quarks are instead electrons with *fractional charges* as long as the electrons orbit the *inside* protons; and *full charges* when the electrons are caught at the outer fringes of the atom, i.e., all the electrons of hydrogen and helium atoms, etc., and those electrons orbiting the *outer protons* of all other atoms would be measured as having their full charge.

The actual mechanism which may therefore determine the density of all material matter, would simply depend upon the numbers of approaches of *exposed outer* protons and outer electrons of each and every atom; so that it would actually come down to the repulsive force which protons have upon protons and electrons upon electrons from one atom to another. This action should attribute to all matter the extent of its density.

For example, the materials comprising more massive atoms like gold or osmium have a much lower frequency of *proton to proton* and *electron to electron repulsive approaches* due to their higher mass numbers resulting in larger diameters and more excess neutrons, thereby keeping the number of repulsive approaches down, and therefore keeping their atoms *closer together* on average. Whereby, with hydrogen, helium and aluminum, for example, the *repulsion frequency* of proton to proton approaches and electron to electron approaches is enormously greater since their mass numbers are so much lower, having few if any excess neutrons to help reduce repulsive approaches. Therefore, the higher the frequency of proton to proton repulsive encounters, the less will be the density of the material in question—the best example, of course, being *Hydrogen gas.*

One other point to be considered along these lines is this: If our present classical structure of atoms is in fact correct, then gold, for example—whose atoms each have 79 electrons supposedly orbiting each nucleus billions of times per millionth of a second—should be the *least dense* of all material matter since each and every gold atom's nucleus would be more or less veiled from each other; while electrons only would always be approaching electrons and therefore create nothing but *repulsive* action between the gold atoms. This would seem to indicate then that the density of gold should be much less than it actually is. If however, a single electron orbits each individual proton as I have advocated, it would naturally be more understandable why gold has, in fact, the great amount of density as it does. It is simply that there would then be much less proton to proton or electron to electron repulsive incidents; and in fact, many more proton-electron attractive incidents between the gold atoms, keeping them closer together and therefore, resulting naturally in the *high* density of gold, per se.

11

Speculations

On the Quantum Riddle, 3°K Radiation, Quasars, and Tides

If in fact all matter throughout the universe were indeed in a constant state of *acceleration,* it then becomes likely that there may exist other explanations for a multitude of phenomena with which we are familiar. A few of these will be discussed in this chapter, but only briefly touched upon. In particular, I will be *highly speculating* in attempting to show that, because I believe all matter is naturally and eternally accelerating at a constant rate of 2.8 Angstroms per second per second, there may be other, more meaningful explanations for phenomena such as: the Evasive Electron and its Momentum-Position riddle; the celebrated 3°K Background Radiation; Quasars; the Earth's Ocean Tides; the Sun's Surface Activities; and another explanation for Isostasy.

On The Quantum Riddle

The Quantum Riddle has been unanswered now for at least the past fifty to sixty years.

The quantum riddle is this: Physicists can determine with *precision* the electron's *position;* they can also determine with *precision* the electron's *momentum* (momentum is the product of a particle's mass and velocity).

The problem, very simply put, is this: physicists *cannot* determine with precision *both,* the electron's position and the electron's momentum *simultaneously.* If they measure one parameter precisely, it has been impossible to measure the other parameter with the same precision at the same precise moment of time.

So there we have it! It's as plain as that. There is however a second part to this riddle. That is the riddle of the *wave-particle dualism.* Again, simply put, the wave-particle riddle is this: Electrons always behave like *waves* when they're not being observed; however, when they *are* being observed, electrons always look like *particles.* It has the characteristic of being both a *wave* and a *particle,* and thus, the other half of the Quantum Riddle. As a *particle,* it appears to be a minuscule point-like mass whose size is approximately 10^{-18} meter in diameter. However, as a *wave,* it is apparent that it is smeared out over large volumes of space which is referred to as the "Airy Pattern" after British Astronomer George Biddell Airy who originally explained this particular phenomena in the early 1800s.

Is it possible that the answer to this riddle can be explained—and thus more *fully understood*—when we take into consideration what the theory in this book, "The New Gravity" is all about?

Does it not seem that, if indeed every particle in the universe is *eternally accelerating* at the constant rate of 2.8 Angstroms per second per second, that once *this factor* is built into the equation it should then be clear *why* the electron appears as a *particle* when it is being observed, and why it has the apparent characteristic of a *wave* when it is not being observed?

Should it not be considered that if a particle with a point-like size of 10^{-18} meter is eternally accelerating at a constant rate of $2.8 \times 10^{-10} \text{ms}^{-2}$, that this would—in reality—mean that electrons are *eternally* increasing their natural basic velocity equal to nearly a *hundred-million* times the 10^{-18} meter length of their diameter for each and every second of time. In other words, every electron allegedly has an eternal and natural built in acceleration of 10^{-18} meter per 10^{-8} second per 10^{-8} second. (This is elaborated upon in more detail in my first book, *2.8 Angstroms*—The Unifying Force of G and c, published in 1990).

The question, once again, is this then: If in fact a particle is *eternally* accelerating due to some *natural* and *eternal* Outside Force acting on it, should it not then be expected to be somewhere else other than where one would expect it to be if this so called eternal acceleration were *not* realized?

If however, the rate of the eternally accelerating electron (and proton, and neutron, etc.) were known, would this not address the question of both, the wave-particle duality and the momentum-position riddle? And would this not also mean that there is no such thing as *constant velocity* in the "particle world?"

On The 3°K Background Radiation

Can it be that the 3°K Blackbody Background Radiation is due directly to our own basic velocity through space of what I postulate to be 300,000 kilometers per second, rather than the *"left over radiation of a so-called big bang?"*

I refer here to the Cosmic Background Radiation detected by Arno Penzias and Robert Wilson of Bell Labs in 1965. This blackbody radiation is usually referred to as the 3°K Background Radiation, but as is known, it is actually closer to 2.7° Kelvin.

What is significant about this background radiation is that it appears to fill all space and is almost exactly the same temperature in all directions. I use the term "almost" because of some measurements made in the late 1970s by Richard Muller and George Smoot which indicated that this background radiation was not as perfectly smooth as was originally believed. Instead, they found it to be about 0.0035°K warmer than average in the direction of the constellation Leo, and at exactly 180° in the opposite direction, toward the constellation Aquarius, it was found to be 0.0035°K cooler than average.

It was also found that this background radiation declined in temperature in a smooth manner as measured from the warm spot in Leo to the cool spot in Aquarius. This *temperature variation* was then properly accounted for as being due to the Earth's overall motion through the Cosmos—that is—we are traveling in the Milky Way from Aquarius toward Leo at a speed of about 390 kilometers per second. Specifically, this speed corresponds to a temperature excess of 0.0035° Kelvin. (Author's Note: these details concerning the cosmic background radiation temperature variation can be found in the text entitled *Universe*, by William J. Kaufman, published 1985, by W. H. Freeman & Co.).

The main point I wish to emphasize is that since this temperature excess of 0.0035°K specifically corresponds to a speed of 390 kilometers per second, then the whole of the 2.7° background radiation we detect can very precisely be equated to what I propose to be the Earth's basic speed through space of 299,792 kilometers per second;

since on the basis of 0.0035°K being equated with a velocity of 390 kilometers per second, then likewise, the velocity of Earth through space of 299,792 kilometers per second precisely equates with a temperature excess of 2.69 degrees Kelvin! Therefore, I would strongly suspect that the total of the 2.7° background radiation may actually be due, "not" to the remnants of a so-called Big Bang, but rather, to the Earth's basic and present velocity of 299,792 kilometers per second through space. For when one calculates these numbers, it will be found that 390 kilometers per second divides into 299,792 kilometers per second 768.7 times. If we then multiply this 768.7-to-one ratio by 0.0035°, we arrive at a product of 2.69° Kelvin.

In other words, the celebrated 2.7° Cosmic Background Radiation may very simply be due to the fact that the Earth is indeed speeding through space with a present velocity of approximately 300,000 kilometers per second, thereby providing us with a more direct, more realistic reason to understand and explain the *real origin* of the Background Radiation.

The satellite COBE (Cosmic Background Explorer) has recently been launched. One of its functions is to find and confirm these *deviations* in the background radiation similar to the existing evidence from the investigations of Smoot & Muller. As far as I am aware, much has indeed already been confirmed at this writing—that is—as relates to the slight variation of temperature. Much more data of course is being received and evaluated from the COBE project which is being headed up by the same George Smoot (who along with Richard Muller) made some similar measurements in the 70s as I have indicated earlier.

On Quasars

Is there a logical explanation for Quasars? Imagine if you will—two hot stars such as the Sun—each rushing toward each other from the outer edges of *two different* expanding universes. Each star then collides head on with the other at a speed of say 300,000 kilometers per second, or possibly even as high as 600,000 km per second, if not more. The energy, and or total light output from two hot stars colliding *head on* at 300,000 kilometers or more per second each, should—in theory—be equal to, or similar to what astrophysicists presently observe when they are looking at Quasars. For if what I have already proposed (as concerns a universal acceleration) are facts to be reckoned with, then there may well be good reason to suspect that Quasars are in fact *two colliding stars* somewhere near the *outer* edges of our own expanding universe and another foreign universe—both of whose outer edges have finally come to expand into each other's turf—something unavoidable in time if more than one universe exists. The energy released from such a collision should be comparable with what is presently observed with Quasars—*a classical example of the energy release according to* $E = mc^2$, thus we would have for (2) colliding stars

$$E = (m_1 \overrightarrow{V^2})\,(m_2 \overleftarrow{V^2}) \qquad (11\text{--}1)$$

On Ocean Tides and Solar Surface Activity

Another interesting speculation concerns the Earth's *Ocean Tides*. For example, what relation might exist between Ocean Tides and Earth's 24-hour rotation as it *continuously accelerates* through space in one general direction? That is, what effect would a 2.8 Angstrom per second per second eternal rate of acceleration upon the Earth have on the waters of the Earth's oceans as each point on Earth swings around to the specific direction of it's basic forward thrust through space, that is, in the opposite direction of t = 0?

Would not the plain fact be that an accelerating body, *containing water,* would simply displace the water from its position of equilibrium? Thus, if the accelerating body is also rotating, would not the displacement of the water be affected at all locations *in time,* with the rate of rotation of the accelerating body—such as is the case with Earth. And when a body of water is displaced at one point on the accelerating body, should not a like displacement take place 180° circumvent of that body of water?

Of course, this all brings into question the Moon and its well known gravitational effect on the Earth. Does the Moon's gravity really play that substantial part of the tide maker as such? Or is it really this other side of the coin— the *acceleration* of all material bodies in the universe at the rate of 2.8 Angstroms per second per second? This would appear to be an interesting area to investigate.

Also, what relation might exist between Solar Prominences and other *solar surface activity* as the Sun itself continuously accelerates through space at 2.8 Angstroms per second per second; rotating as it does, one full revolution every 27 to 31 days?

There are still many unanswered questions concerning Solar Prominences, and if indeed the Sun does basically accelerate through space at 2.8 Angstroms per second per second, then this constant rotation in itself might supposedly explain the Sun's Prominences, just as constant rotation of the Earth (as suggested above) may also explain the *slight prominences* which we refer to as *ocean tides.*

In any case, where is there a massive body in respect to the Sun's position in space which can actually cause such large solar prominences? Neither the Earth's pull of gravity, nor the rest of the planets in the solar system can have the grossly enormous effect upon the Sun such as we observe of its solar prominences. There must then be something else; and a Sun eternally accelerating at 2.8 Angstroms per second per second may just possibly provide that impetus!

One factor which can be explained of the Sun if it is indeed accelerating is that which concerns its loss of mass. It is said that the Sun is losing mass at a rate of four million tons per second. However, if it is indeed accelerating—albeit only 2.8 Angstroms per second per second—then according to $E = mc^2$, lost mass should be replaced since energy and mass are interchangeable. And since c^2, as I propose, is actually the square of a body's velocity through space, then it stands to reason that the Sun's mass is constantly increasing as a result of its constant increase in velocity. Of course, as stated previously, it is constantly losing mass; so that the net effect is that its mass should not actually be decreasing by the large amounts it is presently believed to be.

The Fine Structure Constant

In closing this chapter, I present for further speculation the following equation on the Fine Structure Constant utilizing the Universal Constant of Acceleration a_u for an elapsed time of one-second

$$FSC = (a_u \cdot 1.0 \ sec)^2 (c)^2 \qquad (11-2)$$
$$FSC = (2.795 \times 10^{-10} ms^{-1})^2 (2.998 \times 10^8 ms^{-1})^2$$
$$FSC = 7.021 \times 10^{-3}.$$

Comparing equation (11–2) with the classical equation, we have

$$FSC = \mu_o c e^2 / 2h \tag{11-3}$$
$$FSC = (4\pi \times 10^{-7})(2.998 \times 10^8)(1{,}602 \times 10^{-19})^2 /$$
$$(2.0 \times 6.626 \times 10^{-34})$$
$$FSC = 7.297 \times 10^{-3}.$$

As can be seen, the differences are close enough to give serious consideration to the more simple form in eq. (11-2).

12

GUTs and TOEs

Grand Unified Theories and Theories of Everything! That is what physicists the world over are in search of— GUTs and TOEs—A Grand Unified Theory in which a *single force* is responsible for creating all the forces of nature; a Theory of Everything which they demand must encompass through one set of equations all the important physical phenomena of nature.

This book has purported to promulgate such a theory by having introduced a *redefinition* of Gravitational Mass along with its accompanying *equations* utilizing an Outside Force F_o and its resultant Universal Constant of Acceleration a_u, which force is eternally driving each and every particle throughout the universe at a constant rate of 2.8 Angstroms per second per second.

This Universal Force which creates the Universal Acceleration would therefore mean that it is *not* actually *Light* which possesses the basic energy of motion through

157

space, but in fact, it will show that it is the *material bodies themselves* which possess the *basic motion,* such that, here in our own particular space-time frame, it is we—the Earth and everything on it—that is actually rushing off through space at the enormous speed of 186,282 miles (300,000 km) per second, the speed we attribute to light. Thus, it is for this reason, and this reason alone, that light has the energy to propagate as it does, *mimicking precisely* our own real and basic velocity through space.

If in fact every particle, every body is naturally accelerating, this should then provide the answer as to why we experience a force of attraction at the Earth's surface as we do; so that Gravitation, per se, turns out to be a direct consequence of the fact that all matter is eternally accelerating at a rate of 2.8 Angstroms per second per second; and that the propagation and precise velocity of light are a direct result of the emitting body's velocity through space.

It appears therefore, that we are presently moving thru the universe at the phenomenal speed of 186,282 miles per second after having accelerated for the past 34 billion years of *time* at the rate of 2.8 Angstroms per second per second through a *distance* from time zero of 17 billion *light years*.

As this book, *The New Gravity,* will attempt to make clear, light's energy of propagation and its precise velocity will be shown to be a direct result of our own motion thru space at our present velocity of 186,282 miles per second. This then can be our way of knowing just how fast we are moving thru space. We would, in effect, thereby be able to use light as a *measuring rod* to determine our own precise velocity thru the universe.

Physicists have been questioning for many years whether or not the value of Newton's universal gravitational constant G may be changing with time. The theory in this book postulates that it is *light's velocity* that is in fact changing with time. It is changing by 2.8 Angstroms $(2.8 \times 10^{-10}$ meter) per second per second, or about *one-*

part in *one-thousand quadrillion* for each second of time. That is one part in 10^{18} per second, or one part in 10^{10} per year. They suspect that G may be changing by one part in 10^{10} per year.

GUTs and TOEs demand much from *any* theory that connects the four forces of nature to one single *master* force. The master force of the New Gravity is *alleged* to be some as yet unidentified *Outside Force* F_o which acts upon each and every proton, electron, and planet with just enough force to cause them to accelerate at the same constant rate of $2.8 \times 10^{-10} ms^{-2}$. That constant acceleration is due to the Outside Force of 2.8×10^{-10} Newton acting on a body in *direct proportion* to its column mass m_c—so that for each and every kilogram calculable from the body's column mass, a natural force of 2.8×10^{-10} Newton is applied to the whole body. With the Earth and Moon for example, their total Outside Forces F_o are 9.8 N and 1.6 N respectively—this is a 6-to-1 ratio—just as their column-masses are a 6-to-1 ratio.

It is also because of this *total* Outside Force which pushes a body thru space, that there is an *equal* but *opposite attracting* force *pulling* at all bodies at rest on the surface of the first body; and just as the total Outside Force on a body adds up at the rate of $2.8 \times 10^{-10} N$ for each kilogram of mass in the body's column, so too does the equal but opposite attracting force F_g add up according to each kilogram of mass which rests upon the surface of the first body.

If these same bodies resting at the surface were instead *freely falling* bodies due to gravity near the Earth's surface, they would all fall at the same rate of $9.8 ms^{-2}$, just as *all* bodies throughout the universe accelerate in free space at $2.8 \times 10^{-10} ms^{-2}$.

In closing, there remains one item which deserves to be elaborated upon. That is the *Strong Nuclear Force*—the force which binds Protons and Neutrons together to give stability to the atom.

The Strong Force can be envisioned when one realizes that the g force, that is, the acceleration due to gravity near the surfaces of protons and neutrons (although only about 10^{-7} meter per second per second) is still, to each, an enormous attracting force when one considers that at 10^{-7} meter per second per second, this means that the gravitational pull (the magnitude of g at the surface) of one proton upon one neutron is equivalent to a rate of fall—which when measured in proton *diameters*—would be at a rate of a *million times the proton's diameter per second per second*. However, even of more significance is the fact that the *total* Outside Force F_o acting on the proton and neutron is 10^{-7} Newton.

Compare that to a human body falling to Earth in which the acceleration rate is equal to only six body lengths per second per second. Thus in this case the body accelerates each second only a small minute fraction of the total diameter of the Earth.

If however, we consider two bodies such as one proton and one neutron (instead of a human body and a body the size and mass of Earth), the proton and neutron fall towards each other, so to speak, at a rate equal to approximately *one million* times per second per second their own lengths. In addition, the attracting body they fall towards is the same size and mass as the falling or attracted body itself.

When we are reminded then that the gravitational force is so weak that at the Quantum level it has *no significance* at all, nor nothing whatsoever to do with the Strong Nuclear Force, this would appear to be incorrect; for those quantum level minuscule forces are enormously more powerful on a *proton to neutron scale*, than the attractive forces are on a *large body scale*, such as with planets and people.

So when all is said and done, it may well turn out to be that the Strong Nuclear Force is really the result of this *tiny* force F_o, and its *resultant* eternal energy—the Universal Constant of Acceleration a_u.

It is this *minuscule source of eternal force and resultant energy*—as determined for each body in strict accordance with its *gravitational column mass*—which makes the universe tick; which makes $E = mc^2$ actually $E = mv^2$; which gives the propagation of light its unique quality of a *built in velocity;* which causes the universe to expand as it does; and, amongst several other areas it relates to in physics, it allegedly supplies the answer to the wave-particle dualism and the momentum position riddle of the electron.

So if you ever wondered about the energy which keeps the miraculous pump, *your heart,* working and working and working, you might then be able to get just a *glimpse* of how *God* and *Nature* work, for therein, with the Universal Constant of Acceleration a_u, might lie the very basis of the *energy* source for the *biological* processes as we know them to be.

Einstein, as I had previously stated, may have unwittingly steered the scientific community slightly off course when he postulated an *absolute unchanging* velocity of light; but he was indeed correct about various other matters of great significance which, without many of his good works, the material in this book would not have been possible to promulgate and propound as such. Thus, whatever the message in this book may be worth, none of it, valid or otherwise, could have come about without the labors and genius of other giants of science such as Newton, Galileo, Hubble, the present Allan Sandage, and the many others, past and present, who have made meaningful contributions to science. I also attribute much credit to those present day authors whose books I have studied and cited in the Bibliography, especially those on Quantum mechanics.

Appendix
I

Equations Utilized in This Book

Equations in this book utilizing the Outside Force F_o; the Universal Constant of Acceleration a_u; the Gravitational Mass m_g; the Intertial Mass; the Column Mass m_c; the Force of Gravity F_g; Newton's classical formulations of Gravity and Gravitation; and miscellaneous equations.

Symbol

$m_g = \rho r \cdot 1 m^2$	Gravitational Mass of Sphere
$M = 4\pi/3 \cdot r \cdot r \cdot r \cdot \rho$	Inertial Mass of Sphere
$_u F_o = F_g/m_c$	Universal Outside Force
$m_c = \rho r \cdot 1 m^2$	Column Mass
$m_c = F_g/a_u$	Column Mass
$m_c = F_o/a_u$	Column Mass
$a_u = F_g/m_c$	Universal Constant of Accleration
$a_u = F_o/m_c$	Universal Constant of Acceleration
$a_u = H_o/2$	Universal Constant of Acceleration

$F_o = m_c a_u$	Outside Force
$F_g = m_c a_u$	Force of Gravity at Surface
$a_u = 8\pi/9$	Pure Number (Theoretical Value for a_u)
$G = 8\pi/9 \div 12\pi/9$	Pure Number (Theoretical Value for G)
$G = 8\pi/9 \div 4\pi/3$	Pure Number (Theoretical Value for G)
$V_c = r \cdot 1m^2$	Column Volume of Sphere
$V = 4\pi/3 \cdot r \cdot r \cdot r$	Volume of Sphere
$\rho = M/V$	Density of Sphere (mean)
$a = F/m$	Newton's Second Law
$F = ma$	Newton's Second Law
$G = gr/M_2$	Universal Gravitational Constant (Classical)
$F = GM_1 M_2/r_2$	Universal Law of Gravitation (Classical)
$g = GM/r^2$	Gravitational Field Strength (Classical)
$V_{MW} = a_u t_{MW}$	Velocity of Milky Way
$c_{MW} = a_u t_{MW}$	Milky Way's Velocity of Light
$d_{MW} = (1/2c)t/m$ ly^{-1}	Distance Traveled of Milky Way
$t_{MW} = c/a_u$	Age of Milky Way in Sec.
$A_{MW} = t/sec\ yr^{-1}$	Age of Milky Way in Years
$R = c^2/(a_u \cdot 1sec)^2$	Ratio of Light to Gravity
$FSC = (a_u \cdot 1sec)^2 c^2$	Fine Structure Constant

Appendix II

Constants, Values, Physical Parameters

a_u	Universal Constant of Acceleration	$2.795\ 008 \times 10^{-10} \mathrm{ms}^{-2}$
a_u	Theoretical	$2.792\ 526\ 803\ 16 \times 10^{-10} \mathrm{ms}^{-2}$
$_uF_o$	Universal Outside Force	$2.795\ 008 \times 10^{-10} \mathrm{Nkg}^{-1}$
$_uF_o$	Theoretical	$2.792\ 526\ 803\ 16 \times 10^{-10} \mathrm{Nkg}^{-1}$
F_o	Outside Force (New Gravity)	$2.997\ 924\ 58 \times 10^{8} \mathrm{ms}^{-1}$
c	Present speed of light	$2.997\ 924\ 58 \times 10^{8} \mathrm{ms}^{-1}$
G	Universal Gravitational Constant	$6.672\ 59 \times 10^{-11} \mathrm{m}^3 \mathrm{kg}^{-1} \mathrm{sec}^{-2}$
H_o	Hubble Constant	57.530 km/sec/megaparsec
$H_o/2$	Universal Constant of Acceleration	28.765 km/sec/megaparsec

165

h	Planck Constant	$6.626\ 075\ 5 \times 10^{-34}$ Joule Second
h/e	Planck Constant	$4.135\ 669\ 2 \times 10^{-18}$ keV second
e	Elementary Charge	$1.602\ 177\ 33 \times 10^{-19}$ Coulomb
π	(ρi) Ratio of Circumference/Diameter	$3.141\ 592\ 653\ 59$
$4\pi/3$		$4.188\ 790\ 204\ 77$
$12\pi/9$		$4.188\ 790\ 204\ 77$
$8\pi/9$	a_u's Theoretical Value	$2.792\ 526\ 803\ 16$
$8\pi/9 \div 4\pi/3$	G's Theoretical Value	$0.666\ 666\ 666\ 66.\ .\ .\ .$
$8\pi/9 \div 12\pi/9$	G's Theoretical Value	$0.666\ 666\ 666\ 66.\ .\ .\ .$
$G = a_u/(4\pi/3)$	G's Theoretical Value	$0.666\ 666\ 666\ 66.\ .\ .\ .$
M_{bfs}	Mass of Basic Fundamental Sphere	1.0 Kilogram
$m_{c_{bfs}}$	Column Mass of Basic Fundamental Sphere	1.0 Kilogram
r_{bfs}	Radius of Basic Fundamental Sphere	0.488 602 511 906 m
V_{bfs}	Volume of Basic Fundamental Sphere	$0.488\ 602\ 511\ 906\ \text{m}^3$
ρ_{bfs}	Density of Basic Fundamental Sphere	$2.046\ 653\ 415\ 89\ \text{kg/m}^3$
A_{bfs}	Area of Basic Fundamental Sphere	$3.000\ 000\ 000\ 000\ \text{m}^2$
$F_{g_{bfs}}$	Force of Gravity at Surface of Basic Fundamental Sphere	$2.795\ 008 \times 10^{-10}\text{N}$
$8\pi/9$	(Theoretical)	$2.792\ 526\ 803$ $16 \times 10^{-10}\text{N}$
$F_{g_{bfs}}$	Acceleration Due to Gravity Near Surface of Basic Fundamental Sphere	$2.795\ 008 \times 10^{-10}\text{ms}^{-2}$
$8\pi/9$	(Theoretical)	$2.792\ 526\ 803$ $16 \times 10^{-10}\text{ms}^{-2}$
g_E	Gravity at Earth's Surface	9.806 $6 \times 10^{0}\text{Nkg}^{-1}(\text{ms}^{-2})$

F_g	Gravity at Earth's Surface	$9.806\ 6 \times 10^0 N$
M_E	Mass of Earth	$5.971\ 3 \times 10^{24}$ kg $(5.971\ 3 \times 10^{24})$
ρ_E	Density of Earth (mean)	$5.511\ 8 \times 10^3 kg/m^3$
r_E	Earth Radius	$6.371\ 3\ 10^6$ m
V_E	Earth Volume	$1.803\ 3 \times 10^{21} m^3$

Hubble Constant H_o

57.530 Km Per Sec Per Megaparsec
17.639 Km Per Sec Per Million Light Years
10.960 Miles Per Sec Per Million Light Years
 5.59 Angstroms Per Second Per Light Second

Hubble Constant ($H_o/2$), Adjusted to Account
For The Universal Constant of Acceleration, a_u

28.765 Km Per Sec Per Megaparsec
 8.819 Km Per Sec Per Million Light Years
 5.480 Miles Per Sec Per Million Light Years
 2.795 Angstroms Per Sec Per Light Second

Velocity of Light (present) Milky Way

$2.997\ 924\ 58 \times 10^8$	Meters Per Second
$2.997\ 924\ 58 \times 10^{18}$	Angstroms Per Second
$1.862\ 823\ 96 \times 10^8$	Miles Per Second

Length of One Light Year (present) Milky Way

$9.460\ 528\ 404\ 87 \times 10^{12}$ Kilometers
$5.878\ 499\ 792\ 96 \times 10^{12}$ Miles

| *Megaparsec* | $= 3.261\ 6 \times 10^6$ Light Years |
| *Seconds Per Year* | $= 3.155\ 692\ 597\ 47 \times 10^7$ |

| *Age of Milky Way Galaxy* | $= 33{,}989{,}358{,}053$ Years |
| *Distance of Milky Way Galaxy from ($t = 0$)* | $= 16{,}994{,}679{,}026$ LY |

Appendix
III

169

h	Planck Constant
m_i	Inertial Mass
m_g	Gravitational Mass
F_o	Outside Force (total) Acting on a Body
$_uF_o$	Universal Outside Force
$t=0$	Time-Zero
π	pi
ρ	Mean Density of a Spherical Body
r	Mean Radius of a Spherical Body
d	Diameter of a Spherical Body
N	Newton
kg	Kilogram
m	Meter
m (M)	Mass
E	Energy
F	Force
c	Velocity of Light
t	Time
s	Second
EMF	Electromotive Force
EMR	Electromagnetic Radiation
M_{bfs}	Mass of Basic Fundamental Sphere
$m_{c_{bfs}}$	Column Mass of Basic Fundamental Sphere
r_{bf}	Radius of Basic Fundamental Sphere
V_{bfs}	Volume of Basic Fundamental Sphere
ρ_{bfs}	Density of Basic Fundamental Sphere
A_{bfs}	Area of Basic Fundamental Sphere
$F_{g_{bfs}}$	Force of Gravity at Surface of Basic Fundamental Sphere

Bibliography

Adler, Irving. *Inside the Nucleus.* Signet Books, New York, 1964.

Asimov, Isaac. *Atom.* Dutton (Penguin Books), New York, 1991.

Asimov, Isaac. *The Collapsing Universe.* Pocket Books, New York, 1977.

Asimov, Isaac. *The Neutrino.* Avon Books, New York, 1975.

Asimov, Isaac. *The Subatomic Monster.* Mentor Books, New York, 1985.

Barnett, Lincoln. *The Universe and Dr. Einstein.* William Morrow Co., New York, 1966.

Barrow, John D. & Silk, Joseph. *The Left Hand of Creation.* Basic Books, Inc., New York, 1983.

Barrow, John D. *Theories of Everything*. Ballantine Books, New York, 1992.

Beiser, Arthur. *Physics*. Addison-Wesley Publishing Co., Reading, MA, 1986.

Bergamini, David. *The Universe*. Time-Life Books, New York, 1969.

Bernstein, Jeremy. *Einstein*. Penguin Books, New York, 1985.

Bernstein, Jeremy. *Three Degrees Above Zero*. Mentor Books, New York, 1986.

Berry, Michael. *Principles of Cosmology and Gravitation*. Cambridge University Press, New York, 1978.

Bohm, David. *Quantum Theory*. Dover Publications, Inc., New York, 1989.

Boslough, John. *Stephen Hawking's Universe*. Quill/William Morrow, New York, 1985.

Briggs, John and Peat, F. David. *Turbulant Mirror*. Harper and Row, New York, 1990.

Brooks, William O.; Tracy, George R.; and Tropp, Harry E. *Modern Physical Science*. Holt, Rinehart and Winston, New York, 1962.

Bueche, F. *Principles of Physics*. McGraw Hill Book Co., New York, 1965.

Calder, Nigel. *Einstein's Universe*. Penguin Books, New York, 1980.

Calder, Nigel. *The Key To The Universe*. Penguin Books, New York, 1977.

Capra, Fritjof. *The Tao of Physics*. Bantam Books, New York, 1984.

Capra, Fritjof. *The Turning Point*. Bantam Books, New York, 1988.

Carrigan, Richard A., Jr., and Trower, Peter. *Particles and Forces*. W. H. Freeman and Co., New York, 1990.

Charon, Jean. *Cosmology, Theories of the Universe.* McGraw Hill, New York, 1975.

Clark, Ronald W. *Einstein, The Life and Times.* Avon Books, New York, 1972.

Cole, K. C. *Sympathetic Vibrations.* Bantam Books, New York, 1985.

Coleman, James A. *Modern Theories of the Universe.* Signet Books, New York, 1963.

Cooper, Necia G. and West, Geoffrey B. *Particle Physics.* Cambridge University Press, New York, 1988.

Crease, Robert P. and Mann, Charles C. *The Second Creation.* Collier Books/MacMillen, New York, 1983.

D'Abro, A. *The Rise of the New Physics.* Dover Publications, New York, 1952.

Davies, Paul and Gribbin, John. *The Matter Myth.* Simon & Schuster/Touchstone, New York, 1992.

Davies, P. C. W. *Space and Time in the Modern Universe.* Cambridge University Press, New York, 1977.

Ditfurth, Holmar. *Von. Children of the Universe.* Atheneum, New York, 1974.

Ebbighausen, E. G. *Astronomy.* Charles E. Merrill Books, Columbus, Ohio, 1966.

Einstein, Albert. *Essays In Physics.* Philosophical Library, New York, 1950.

Einstein, Albert. *Letters To Solovine.* Philosophical Library, New York, 1986.

Einstein, Albert. *Relativity.* Crown Publishers, New York, 1952.

Einstein, Albert. *Sidelights On Relativity.* Dover Publications, New York, 1983.

Epstein, Lewis Carroll. *Relativity Visualized.* Insight Press, San Francisco, 1985.

Ferris, Timothy. *Galaxies.* Stewart Tabori & Chang Pub., New York, 1982.

Ferris, Timothy. *The Red Limit.* Bantam Books, New York, 1977.

Feynman, Richard P. *QED—The Strange Story of Light and Matter.* Princeton University Books, New Jersey, 1985.

Feynman, Richard P. *Surely You're Joking Mr. Feynman!* Bantam Books, New York, 1989.

Feynman, Richard P. *The Feynman Lectures On Physics.* (Richard Feynman, Robert B. Leighton and Matthew L. Sands.) Originally published in 1963 by Addision-Wesley, Reading, MA. This Commemorative Three-Volume Issue published by Allan M. Wylde, 1989.

Feynman, Richard P. *"What Do You Care What Other People Think?"* Bantam Books, New York, 1989.

Field, George B.; Arp, Halton; Balecall, John N. *The Redshift Controversy.* W. A. Benjamin, Inc., Reading, MA, 1976.

Fritzsch, Harold. *Quarks.* Basic Books, A Division of Harper Collins, 1983.

Gamow, George. *Gravity.* Anchor Books, Doubleday & Co., New York, 1962.

Gamow, George. *One Two Three—Infinity.* Bantam Books, New York, 1967.

Gamow, George. *Thirty Years That Shook Physics.* Dover, New York, 1966.

Gardner, Martin. *The Relativity Explosion.* Vintage Books, New York, 1976.

Gartenhaus, Soloman. *Physics-Basic Principles.* Holt, Rinehart and Winston, New York, 1975.

Giancoli, Douglas C. *The Ideas of Physics.* Harcourt, Brace Jovanovich, Orlando, FL, 1986.

Gibbins, Peter. *Particles and Paradoxes*. Cambridge University Press, New York, 1987.

Ginzburg, V. L. *Key Problems of Physics and Astrophysics*. Mir Publishers, Moscow, 1978.

Glashow, Sheldon L. *Interactions*. Warner Books, New York, 1988.

Glashow, Sheldon L. *The Charm of Physics*. American Institute of Physics, New York, 1991.

Gleick, James. Chaos. *Penguin Books*, New York, 1987.

Gray, Reginald Irvin. *Unified Physics*. Naval Surface Warfare Center, Dahlgren, VA, 1988.

Gribbin, John. *In Search of The Double Helix*. Bantam Books, New York, 1987.

Gribbin, John and Rees, Martin. *Cosmic Coincidences*. Bantam Books, New York, 1989.

Gribbin, John. *In Search of The Big Bang*. Bantam Books, New York, 1986.

Gribbin, John. *In Search of Schrodinger's Cat*. Bantam Books, New York, 1984.

Gribbin, John. *The Omega Point*. Bantam Books, New York, 1988.

Halliday, David and Resnick, Robert. *Physics for Students of Science and Engineering*. John Wiley and Sons, Inc., New York, 1963.

Halliday, David and Resnick, Robert. *Fundamentals of Physics*. John Wiley and Sons, Inc., New York, 1974.

Han, M. Y. *The Secret Life of Quanta*. TAB Books, Blue Ridge Summit, PA, 1990.

Hazen, Robert M. and Trefil, James. *Science Matters*. Doubleday, New York, 1991.

Herbert, Nick. *Quantum Reality*. An Anchor Book published by: Doubleday, New York, 1987.

Hey, Tony and Walters, Patrick. *The Quantum Universe.* Cambridge University Press, New York, 1987.

Hoffmann, Banesh. *The Strange Story of the Quantum.* Dover Publications, New York, 1959.

Hooper, Henry O. and Gwyne, Peter. *Physics and the Physical Perspective.* Harper and Row, New York, 1977.

Hoyle, Fred. *Ten Faces of the Universe.* W. H. Freeman and Co., San Francisco, 1977.

Jean, Sir James. *The Growth of Physical Sciences.* Fawcett Publications, Greenwich, CT, 1967.

John, Laurie. *Cosmology Now.* Taplinger Publishing Co., New York, 1976.

Kaku, Michio and Trainer, Jennifer. *Beyond Einstein.* Bantam Books, New York, 1987.

Kaufmann, William J. III. *Relativity and Cosmology.* Harper and Row, New York, 1977.

Kaufmann, William J. III. *Universe.* W. H. Freeman and Co., New York, 1985.

Krauskopf, Konrad B. and Belser, Arthur. *The Physical Universe.* McGraw-Hill Book Co., New York, 1976.

Landau, L. D. and Kitaigorodsky, A. *Physical Bodies.* Mir Publishers, Moscow, 1980.

Landau, L. and Rumer, Yu. *What is the Theory of Relativity.* Mir Publishers, Moscow, 1974.

Lederman, Leon with Teresi, Dick. *The God Particle.* Houghton Mifflin Co., New York and Boston, 1993.

Levitt, I. M. *Beyond the Known Universe.* The Viking Press, New York, 1974.

Logunov, A. A. *Gravitation and Elemental Particle Physics.* Mir Publishers, Moscow, 1983.

Long, Dale D. *The Physics Around You.* Wadsworth Publishing Co., Belmont, CA, 1988.

Lorentz, H. A.; Einstein, Albert; Minkowski, H.; Wayl, H. *The Principle of Relativity*. Dover Publications, New York, 1952.

Lusanna, L. *New Trends In Particle Theory*. World Scientific Pub., Singapore, 1985.

McAleer, Neil. *The Mind Boggling Universe*. Doubleday and Co., New York, 1987.

Maxwell, James Clerk. *Matter and Motion*. Dover Publications, New York, 1991.

Menzel, Donald H.; Whipple, Fred; L., Vaucouleurs, Gerard de. *Survey of the Universe*. Prentice-Hall, Englewood Cliff, New Jersey, 1970.

Miller, Franklin, Jr. *College Physics*. Harcourt, Brace and World, Inc., New York, 1967.

Miller, Franklin Jr. and Schroeer, Dietrich. *College Physics*. Harcourt, Brace Jovanovich, Orlando, FL, 1959.

Monk, George S. *Light-Principles and Experiments*. Dover Publications, Inc., New York, 1963.

Morris, Richard. *The Edges of Science*. Prentice Hall Press, New York, 1990.

Morris, Richard. *The Nature of Reality*. The Noonday Press, New York, 1987.

Oppenheimer, Robert J. *Atom and Void*. Princeton University Press, Princeton, New Jersey, 1989.

Pagels, Heinz R. *Perfect Symmetry*. Bantam Books, New York, 1986.

Pagels, Heinz R. *The Cosmic Code*. Bantam Books, New York, 1984.

Pais, Abraham. *Inward Bound*. Oxford University Press, New York, 1986.

Pais, Abraham. *Subtle is the Lord*. Oxford University Press, New York, 1982.

Parker, Barry. *Einstein's Dream.* Plenum Press, New York, 1987.

Parker, Barry. *Search for a Supertheory.* Plenum Press, New York, 1987.

Pasachoff, Jay M. and Kuther, Marc L. *Invitation to Physics.* W. W. Norton and Co., New York, 1981.

Peat, F. David. *Einstein's Moon.* Contemporary Books, Inc., Chicago, 1990.

Physical Science Study Committee. *Physics.* D. C. Heath and Company, Boston, 1965.

Plutchik, Robert. *Foundations of Experimental Research.* Harper and Row, New York, 1968.

Preiss, Byron-Editor. *The Microverse.* Bantam Books, New York, 1989.

Reid, R. W. *The Spectroscope.* Signet, New York, 1966.

Riordan, Michael. *The Hunting of the Quark.* Simon and Schuster, New York, 1987.

Riordan, Michael and Schramm, David N. *The Shadows of Creation.* W. H. Freeman and Co., New York, 1990.

Rowan-Robinson, Michael. *The Cosmological Distance Ladder.* W. H. Freeman and Co., New York, 1985.

Russell, Bertrand. *The ABC of Relativity.* Signet Books, New York, 1958.

Sagan, Carl. *Cosmos.* Random House, New York, 1980.

Schatzman, E. L. *The Structure of the Universe.* McGraw-Hill, New York, 1976.

Sears, Francis W.; Zemansky, Mark W.; and Young, Hugh D. *College Physics.* Addison-Wesley Publishing Co., Reading, MA, 1986.

Segre, Emilio. *From X-Ray to Quarks.* W. H. Freeman and Co., New York, 1980.

Serway, Raymond A. and Faughn, Jerry S. *College Physics.* Saunders College Publishing, New York, 1985.

Singh, Jagjit. *Great Ideas and Theories of Modern Cosmology.* Dover Publications, New York, 1970.

Smorodinsky, Ya A. *Particles, Quanta, Waves.* Mir Publications, Moscow, 1976.

Stewart, Ian. *Does God Play Dice.* Blackwell Publishers, Cambridge, MA, 1992.

Tilley, Donald E. and Thumm, Walter. *College Physics.* Cummings Publishing Co., Menlo Park, CA, 1971.

Toulmin, Stephen and Goodfield, June. *The Fabric of the Heavens.* Harper Torchbook Edition, Harper and Row, New York, 1965.

Trefil, James S. *From Atoms to Quarks.* Scrubners, New York, 1980.

Trefil, James. *The Dark Side of the Universe.* Anchor Books/ Doubleday, New York, 1988.

Watkins, Peter. *Story of the W and Z.* Cambridge University Press, New York, 1986.

Weinberg, Steven. *Dreams of a Final Theory.* Pantheon Books, New York, 1992.

Weinberg, Steven. *The First Three Minutes.* Basic Books, New York, 1977.

White, Harvey E. *Modern College Physics.* D. Van Nostrand Co., New York, 1972.

Whipple, Fred L. *Earth, Moon, and Planets.* Harvard University Press, Cambridge, 1971.

Wilczek, Frank and Devine, Betsy. *Longing for the Harmonies.* W. W. Norton and Co., New York, 1988.

Will, Clifford M. *Was Einstein Right?* Basic Books, New York, 1986.

Williams, John E.; Trinklein, Frederick E.; and Metcalfe, Clark H. *Modern Physics*. Holt, Rinehart and Winston, New York, 1984.

Wilson, Jerry D. *Physics-Concepts and Applications*. D. C. Heath and Co., Lexington, MA, 1977.

Zee, A. *Fearful Symmetry*. Collier Books, Macmillan Pub. Co., New York, 1986.

Zukav, Gary. *The Dancing Wu Li Masters*. Bantam Books, New York, 1979.

About the Author

Kenneth Salem has been working since 1968 to convince "Fluid Dynamacists" that Turbulence in Fluid Flows is the result of "Ordinary Acoustical pressures." In the process of his experimental work he discovered that individual sub-micron size particles, three-tenth micron and *smaller,* can be detected with the naked eye. Upon acquiring a U.S. Patent for an apparatus to bring into play the particle phenomenon he discovered, he had the honor of displaying the naked eye "Particle Detector" to the public at the premier National Inventors Day Exposition held by the U.S. Patent Office in Washington, D.C. Other participants in the show included such prestigious companies as AT&T, Polaroid Corp., and Mazda Corp. with their Wankel Engine.

Salem lives with his wife, Jean, in the historical flood renowned city of Johnstown, Pennsylvania.

Kenneth George Salem—theoretical physicist was honored in the 1992–1993 Premier Edition of *Who's Who in Science and Engineering* (a Marquis Who's Who in America Publication). Some of his achievements include: Author of a Physics Book *2.8 Angstroms*—1990; Articles to professional journals including Jour. Brit.-Am. Sci. Rsch. Assn., Bull. Pure and Applied Sci. Reporting. Patents relating to discoveries that sub-micron particles are detectable with the naked eye; improving flashlight to emit diffuse (shadow free) light; that turbulence in fluid flows is primarily due to acoustical radiation and research on an improved (direct and dimensionally correct) Universal Constant of Gravitation.

SALEM INVENTIONS PATENTED

Top physicists still can't believe that Ken Salem's pollution light works.

The light (*U. S. Pat. No. 3726593) permits naked-eye detection of sub-micron particles in the air. When you consider that a micron is only one-thousandth of a millimeter and there are 25 millimeters to an inch, it's easy to see why scientists were convinced naked-eye detection was impossible.

The Salem Pollution Light could have application in checking dust levels in coal mines or it may be useful for checking clean room status or perhaps for detecting gas or fluid leaks.

Ken, a salesman in the Industrial Department whose customers know he's a bulldog on tracking down orders, does much of his inventing in the den of his Bedford Street home and occasionally he uses the facilities of the electronics lab at UPJ.

The Franklin Borough High School graduate also received a patent for Methods and Apparatus for Visually Illustrating Sound Waves. Ken believes this invention may be his most significant, to date, as it tends to disprove some laws of fluid dynamics. Watching the rabbit ears on his new color TV bounce in time to music triggered his thinking on this invention.

Salem's hobby as an inventor is a good example of the caliber of men representing M. Glosser and Sons. If Ken can't solve your problem, maybe he'll invent an answer.

NATIONAL INVENTORS' DAY

A joint resolution passed by both Houses of Congress, and a proclamation issued by President Nixon designated February 11 as National Inventors' Day in recognition of the valuable contribution of inventors to the economic growth and technological development of the United States.

In Honor of the occasion the Patent Office held an exposition on Sunday, February 11, 1973.

Mr. Salem was one of the inventors invited to this expo to demonstrate how his pollution light worked.

On the following page are pictures taken at Mr. Salem's booth at the inventors' expo.

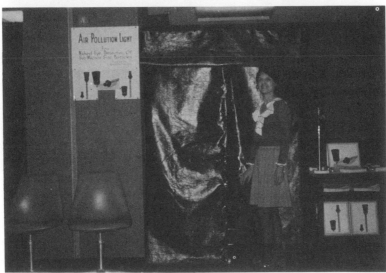

The Salem's at the Kennedy Center (with the Watergate complex in the background) in Washington, D.C., after the National Inventors' Day Expo in February 1973.

Kenneth G. Salem
. . . *his latest invention*

To Be at Arena

Salem Improves On Flashlight

Kenneth G. Salem of 226 Belmont St. is an inventor who believes in the adage about a better mousetrap.

But for him, it's an improved flashlight that may cause the world to beat a path to his door. His brainchild will be displayed at the Greater Johnstown Industrial Exposition, which opens Thursday at the Cambria County War Memorial Arena.

A patent application for his improved flashlight is sprinkled liberally with such terms as index of refraction and light intensity. But for the average layman, it suffices to say that the flashlight utilizes several

KEN-LITE™ Ⓖ

novel concepts to produce a beam of light three to four times brighter than normal and with none of the shadow effects inherent in standard flashlights.

Marketing Set

M. Glosser and Sons, Inc., by whom Mr. Salem is employed, has agreed to market the flashlight under the trade name Ken-Lite. The company sees a bright future for the light in industrial, law enforcement, fire protection and mine safety applications, as well as everyday household use. Its inventor already is studying possible applications of his concept to such things as automobile headlights and desk lamps.

Daniel S. Glosser, president of M. Glosser, said he welcomed the Ken-Lite as a valuable addition to the 75-year-old firm's line of industrial supplies.

Mr. Salem conceived the idea for the new flashlight (patent pending) while doing research for the Salem Pollution Light. That invention, patented about a year ago, is being used by a Florida pollution-control company as a visual aid for detecting submicron particles in gas and fluids. Particles of submicron size ordinarily are not visible to the naked eye.

THE NEW GRAVITY may be ordered direct from the publisher. It may also be found in any major American college and University bookstore.

To order direct from the publisher, send $12.00 plus $3.00 shipping and handling.* Send only check or money order. For Canadian residents send $14.00 plus $3.00 shipping and handling.

You may use this order form for ordering:

SALEM Books
P.O. Box 908
Johnstown, PA 15907

Please send me____copies of "THE NEW GRAVITY", ISBN 0-9625398-1-3. I enclose a check or money order for the total amount due.

Name: _____

Address: _____

City: _____

State: _____ Zip: _____

*Pennsylvania residents add applicable sales tax.